STUDY GUIDE
With Selected Solutions for Weltman and Perez's Intermediate Algebra
Second Edition

Vicky Lymbery

Ellen T. Wood

Stephen F. Austin State University

Wadsworth Publishing Company
Belmont, California
A Division of Wadsworth, Inc.

Printed in the United States of America 49

1 2 3 4 5 6 7 8 9 10—94 93 92 91 90

PREFACE

This study guide was written as a supplement to *Intermediate Algebra, Second Edition,* by Weltman and Perez. The method of presentation and the development of material parallels the approach of the parent textbook. The primary purpose in writing this study guide has been to provide review and reinforcement of the material presented throughout the textbook.

Other objectives include:

1. The provision of a variety of types of problems to keep student interest alive.
2. The use of problem-solving methods found in the textbook to provide a correlation between the two books. Many of the odd-numbered exercises from the textbook have been solved in detail. These problem numbers are identified by a box; i.e., $\boxed{3}$.
3. The provision of enough work space for the student to solve each problem.
4. The presentation of a step-by-step complete solution of each problem for the student. The solution page is a study guide in itself.

We wish to express our appreciation to our families, our friends, and our colleagues for their encouragement and support throughout this endeavor. Our sincerest thanks go to reviewers Beverly Abila of Rio Hondo College and Alice K. Hagood of Alvin Community College, whose thoughtful suggestions added much to this book. In addition, we wish to thank Anne Scanlan-Rohrer, assistant editor, and Ragu Raghavan, field editor, of Wadsworth Publishing Company for their assistance.

TO THE STUDENT

This study guide was written as a supplement to the textbook *Intermediate Algebra* by Weltman and Perez. The purpose of this study guide is to give you the most thorough review and reinforcement of the concepts presented in the textbook that is possible.

In this study guide you will find:

1. Many types of problems designed to help you master the basic concepts.
2. Problems taken directly from the textbook with solutions. These problem numbers are identified by a box; i.e., [3].
3. Plenty of work space so that you can work the problems directly on the page.
4. A complete step-by-step solution for each problem.

It is the intention of the authors that you work each problem in the space provided and then check your solution against the solution provided for you. Remember that there is no substitute for the experience gained by actually working through the solution for each problem yourself.

CONTENTS

STUDY GUIDE

With Selected Solutions for Weltman and Perez's Intermediate Algebra

Second Edition

C H A P T E R

1 Fundamental Concepts

1 · 1 Sets

1. A collection of objects is called a ___set___.

2. The numbers or things within a set are called the ___elements___ or the ___members___ of the set.

3. The symbol for "is an element of" is ___$\in$___.

4. If each element of set B is also an element of set A, then set B is a ___subset___ of set A.

5. The symbol for "is a subset of" is ___$\subseteq$___.

6. List the subsets of the set A = {1, 2, 3}.

   $\{1\}, \{2\}, \{3\}, \{1, 2\}, \{1, 3\}, \{2, 3\}, \varnothing, \{1, 2, 3\}$

7. Remember that every set is a ___subset___ of itself and that the ___null___ set or the ___empty___ set is a subset of every set.

C H A P T E R

1 Fundamental Concepts

1·1 Sets

1. A collection of objects is called a ___set___.

2. The numbers or things within a set are called the ___elements___ or the ___members___ of the set.

3. The symbol for "is an element of" is ___$\in$___.

4. If each element of set B is also an element of set A, then set B is a ___subset___ of set A.

5. The symbol for "is a subset of" is ___$\subseteq$___.

6. List the subsets of the set A = {1, 2, 3}.

   {1} {2} {3} {1, 2} {1, 3} {2, 3} ∅ {1, 2, 3}

7. Remember that every set is a ___subset___ of itself and that the ___null___ set or the ___empty___ set is a subset of every set.

Consider the set $A = \{2, 4, 6, \ldots\}$ and the set $B = \{2, 4, 6\}$ and answer TRUE or FALSE for each of the following:

true 8. Set A is an infinite set.

false 9. Sets A and B are equal sets.

true 10. $A \cup B = \{2, 4, 6, \ldots\}$

false 11. $A \cap B = \{2, 4, 6, \ldots\}$

true 12. Sets A and B have elements in common.

true 13. $2 \in A$

false 14. $2 \subseteq A$

false 15. $A \subseteq B$

true 16. $\emptyset \subseteq B$

false 17. $\{4\} \in B$

Consider the sets W = {red, blue, green}, X = {red, yellow, pink}, $Y = \{1, 2, 3, 4, 5, \ldots, 10\}$, and $Z = \{1, 3, 5, \ldots\}$ and give the answers to the following:

18. $W \cap Z = \emptyset$

19. $W \cup X =$ {red, blue, green, yellow, pink}

20. $Z \cap Y = \{1, 3, 5, 7, 9\}$

21. $(Y \cup Z) \cap Y = Y$ or $\{1, 2, 3 \ldots, 10\}$

22. $W \cap X =$ {red}

1 · 2 Natural Numbers and Integers

1. Explain the difference between the set of Natural numbers, N, and the set of Integers, I. The set of Natural numbers includes the counting numbers. The set of Integers includes the Natural numbers, the negatives of the Natural numbers, and zero.

Consider the set A = {2, 4, 6, ...} and the set B = {2, 4, 6} and answer TRUE or FALSE for each of the following:

True 8. Set A is an infinite set.

false 9. Sets A and B are equal sets.

true 10. $A \cup B = \{2, 4, 6, ...\}$ (intersect)

false 11. $A \cap B = \{2, 4, 6, ...\}$ (union)

TRue 12. Sets A and B have elements in common.

TRue 13. $2 \in A$

True 14. $2 \subseteq A$ why? false

false 15. $A \subseteq B$

True 16. $\emptyset \subseteq B$

true 17. $\{4\} \in B$ false why?

Consider the sets W = {red, blue, green}, X = {red, yellow, pink}, Y = {1, 2, 3, 4, 5, ..., 10}, and Z = {1, 3, 5, ...} and give the answers to the following:

18. $W \cap Z$ = $\emptyset$

19. $W \cup X$ = red, blue, yellow, green, pink

20. $Z \cap Y$ = {1, 3, 5, 7, 9}

21. $(Y \cup Z) \cap Y$ = Y

22. $W \cap X$ = {red}

1·2 Natural Numbers and Integers

1. Explain the difference between the set of Natural numbers, N, and the set of Integers, I.

   N = {1, 2, 3, 4, ...}
   I = {... -1, 0, 1, 2 ...}

2. State the rule for adding two integers having opposite signs.

Subtract the smaller absolute value from the larger absolute value. The answer will have the same sign as that of the number with the larger absolute value.

3. State the rule for subtraction of one integer from another integer.

To subtract an integer, add its additive inverse.

$3-(-8) = 3+(8) = 11$

4. In the expression $8 + (-12) = -4$, 8 and (-12) are called terms, and -4 is called the sum.

5. In the expression $(-15) \div 3 = -5$, (-15) is called the dividend, 3 is called the divisor, and -5 is called the quotient.

6. In the expression $(-3) \cdot (-4) = 12$, (-3) and (-4) are called factors, and 12 is called the product.

7. In the expression 3^6, the 3 is called the base and the 6 is called the power or exponent.

8. (a) 5^3 is read "five cubed."

(b) 8^2 is read "eight squared."

9. What is the sign of the product of two integers having opposite signs?

negative

10. What is the sign of the quotient of two integers having the same signs? positive

Perform if possible the indicated operations.

11. $-30 - (-52) =$ 22

12. $-9 - 15 - (-31) =$ 7

13. $(-2)(-1)(15)(-4)(-5) =$ 600

14. $(-3)^5 =$ -243

15. $(-2^4) =$ -16

16. $-3^2 =$ -9

17. $8/0 =$ undefined

18. $0/(-13) =$ 0

19. $-57 + 17 - (-23) - 7 =$ -24

20. $(-3)(-2)(-5) =$ -30

2. State the rule for adding two integers having opposite signs.

subtract smaller from larger then give answer the sign of larger #

3. State the rule for subtraction of one integer from another integer.

add additive INVERSE
$3-(-8) = 3+8 = 11$

4. In the expression $8 + (-12) = -4$, 8 and (-12) are called Terms, and -4 is called the sum.

5. In the expression $(-15) \div 3 = -5$, (-15) is called the dividend, 3 is called the divisor, and -5 is called the Quotient.

6. In the expression $(-3) \cdot (-4) = 12$, (-3) and (-4) are called factors, and 12 is called the product.

7. In the expression 3^6, the 3 is called the base and the 6 is called the power or exponent.

8. (a) 5^3 is read "five cubed."

(b) 8^2 is read "eight squared."

9. What is the sign of the product of two integers having opposite signs?

negative

10. What is the sign of the quotient of two integers having the same signs? positive

Perform if possible the indicated operations.

11. $-30 - (-52) =$ 22

12. $-9 - 15 - (-31) =$ 7

13. $(-2)(-1)(15)(-4)(-5) =$ 600

14. $(-3)^5 =$ -243

15. $(-2^4) =$ -16 (4 inside)

16. $-3^2 =$ -9

17. $8/0 =$ und.

18. $0/(-13) =$ 0

19. $-57 + 17 - (-23) - 7 =$ -24 (-40, -17)

20. $(-3)(-2)(-5) =$ -30

1 · 3 Rational Numbers

1. Circle the rational numbers: { (1/2), (-5), 3/0 , (6.2), (.8), (4) }

2. Find the prime factorizations of 36 and 132.

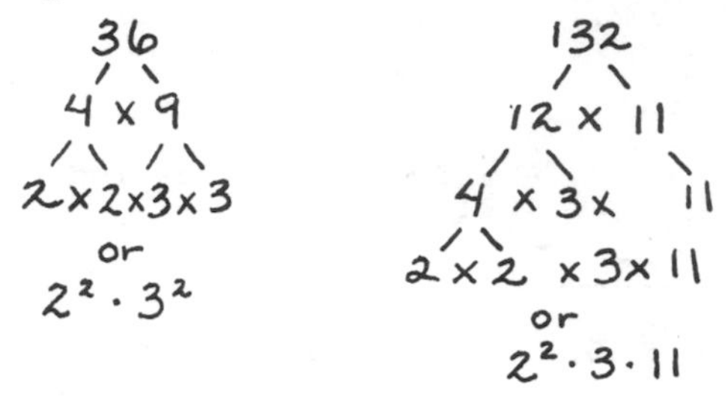

3. Reduce 36/132 to lowest terms.

$$\frac{36}{132} = \frac{\not{2}\cdot\not{2}\cdot\not{3}\cdot 3}{\not{2}\cdot\not{2}\cdot\not{3}\cdot 11} = \boxed{\frac{3}{11}}$$

4. Is -3/4 equivalent to 3/(-4)?

$\boxed{\text{yes}}$ because $-3 \div 4 = -.75$
$3 \div -4 = -.75$

5. Find the Least Common Denominator (LCD) of 36 and 132.

36 = $2^2 \cdot 3^2$ (from problem 2 above)

132 = $2^2 \cdot 3 \cdot 11$

The highest power of 2 is <u>2</u>.

The highest power of <u>3</u> is <u>2</u>.

The highest power of <u>11</u> is <u>1</u>.

Therefore, the LCD is <u>$2^2 \cdot 3^2 \cdot 11^1$</u> = <u>396</u>.

6. Using the above method, find the LCD of 7, 21, and 9.

$7 = 7^1$ The highest power of 7 is 1
$21 = 7^1 \cdot 3^1$ The highest power of 3 is 2
$9 = 3^2$ The LCD is $7^1 \cdot 3^2 = \boxed{63}$

1 · 3 Rational Numbers

1. Circle the rational numbers: { 1/2 , -5 , 3/0 , 6.2 , .8 , 4 }

2. Find the prime factorizations of 36 and 132.

   36: $4 \cdot 9$, $2 \cdot 2 \cdot 3 \cdot 3$ or $2^2 \cdot 3^2$

   132: $12 \cdot 11$, $3 \cdot 2 \cdot 2 \cdot 11$ or $2^2 \cdot 3 \cdot 11$

3. Reduce 36/132 to lowest terms. $\frac{36}{132} = \frac{3}{11}$

4. Is -3/4 equivalent to 3/(-4)? yes

5. Find the Least Common Denominator (LCD) of 36 and 132.

   36 = $2^2 \cdot 3^2$

   132 = $2^2 \cdot 3 \cdot 11$

   The highest power of 2 is 2.

   The highest power of 3 is 2.

   The highest power of 11 is 1.

   Therefore, the LCD is $2^2 \cdot 3^2 \cdot 11$ = 396.

6. Using the above method, find the LCD of 7, 21, and 9.

   $7 = 7 \cdot 1$, $7 \cdot 1$, 7^1

   $21 = 7 \cdot 3$, $7 \cdot 3$

   $9 = 3 \cdot 3$, 3^2, 3^2

   $3 \cdot 3 \cdot 7 = 63$

7. Using the above method, find the LCD of 30, 450, and 24.

$30 = 2^1 \cdot 3^1 \cdot 5^1$ The highest power of 2 is 3

$450 = 2^1 \cdot 3^2 \cdot 5^2$ The highest power of 3 is 2

$24 = 2^3 \cdot 3$ The highest power of 5 is 2

The LCD is $2^3 \cdot 3^2 \cdot 5^2 = \boxed{1800}$

8. 21/60 - (-47/90)

STEP 1: FIND LCD

$60 = 2^2 \cdot 3 \cdot 5^1$

$90 = 2 \cdot 3^2 \cdot 5^1$

$\text{LCD} = 2^2 \cdot 3^2 \cdot 5^1 = 180$

STEP 2: SOLVE

$\frac{21}{60} - \left(\frac{-47}{90}\right) =$

$\frac{21}{60} \cdot \frac{3}{3} - \left(\frac{-47}{90}\right) \cdot \frac{2}{2} = \frac{63}{180} + \frac{94}{180} = \boxed{\frac{157}{180}}$

we are multiplying by the numbers necessary to yield 180 in the denominator.

9. 11/126 + 7/60 - 5/14

STEP 1: FIND LCD

$126 = 2^1 \cdot 3^2 \cdot 7^1$

$60 = 2^2 \cdot 3^1 \cdot 5^1$

$14 = 2^1 \cdot 7^1$

$\text{LCD} = 2^2 \cdot 3^2 \cdot 5^1 \cdot 7^1 = 1260$

STEP 2: SOLVE

$\frac{11}{126} + \frac{7}{60} - \frac{5}{14} =$

$\left(\frac{11}{126} \cdot \frac{10}{10}\right) + \left(\frac{7}{60} \cdot \frac{21}{21}\right) - \left(\frac{5}{14} \cdot \frac{90}{90}\right) =$

$\frac{110}{1260} + \frac{147}{1260} - \frac{450}{1260} = \boxed{\frac{-193}{1260}}$

10. Perform the indicated operations:

(a) (3/4) (-12/21)

$\frac{3}{4} \cdot \frac{-12}{21} = \frac{3 \cdot (-12)}{4 \cdot 21} =$

$\frac{3 \cdot (-\cancel{2})^{-1} \cdot \cancel{2} \cdot \cancel{3}}{\cancel{2} \cdot \cancel{2} \cdot \cancel{3} \cdot 7} = \boxed{\frac{-3}{7}}$

(b) [17] 2/15 ÷ 10/27

$\frac{2}{15} \div \frac{10}{27} = \frac{2}{15} \cdot \frac{27}{10} = \frac{2 \cdot 27}{15 \cdot 10}$

$= \frac{\cancel{2} \cdot \cancel{3} \cdot 3 \cdot 3}{\cancel{3} \cdot 5 \cdot \cancel{2} \cdot 5} = \boxed{\frac{9}{25}}$

(c) [25] (3 1/4) (48/-13) (-5/-8)

$3\frac{1}{4} \cdot \frac{48}{-13} \cdot \frac{-5}{-8} = \frac{13}{4} \cdot \frac{48}{-13} \cdot \frac{-5}{-8}$

$= \frac{-\cancel{13} \cdot \cancel{2} \cdot \cancel{2} \cdot \cancel{2} \cdot \cancel{2} \cdot 3 \cdot 5}{\cancel{2} \cdot \cancel{2} \cdot \cancel{13} \cdot \cancel{2} \cdot \cancel{2} \cdot 2} = \frac{-15}{2}$ or $\boxed{-7\frac{1}{2}}$

7. Using the above method, find the LCD of 30, 450, and 24.

8. 21/60 - (-47/90)

STEP 1: FIND LCD

60 =

90 =

LCD =

STEP 2: SOLVE

9. 11/126 + 7/60 - 5/14

STEP 1: ______________

STEP 2: ______________

10. Perform the indicated operations:

(a) (3/4)(-12/21)

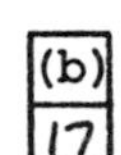
2/15 ÷ 10/27

(3 1/4)(48/-13)(-5/-8)

11. Convert to decimals:

(a) [33] $-2/9 = -2 \div 9 = \boxed{-.2\overline{2}}$

(b) $5/6 = 5 \div 6 = \boxed{.83\overline{3}}$

12. Convert to fractions in lowest terms:

(a) [39] $3.875 = 3\frac{875}{1000}$

$= 3\frac{\not{5}\cdot\not{5}\cdot\not{5}\cdot 7}{2\cdot\not{5}\cdot 2\cdot\not{5}\cdot 2\cdot\not{5}}$

$= 3\frac{7}{8}$ or $\boxed{\frac{31}{8}}$

(b) $.003 = \boxed{\frac{3}{1000}}$

13. Perform the indicated operations:

(a) [45] $(-13.356) \div (-6.3)$

$(-13.356) \div (-6.3) = \boxed{2.12}$

(b) $26.3 - 41.8 + 7.005$

$(26.3 - 41.8) + 7.005 =$
$-15.5 + 7.005 = \boxed{-8.495}$

(c) $(-.0021)(6.01) = \boxed{-.012621}$

14. A rational number is one whose decimal representation either <u>repeats</u> or <u>terminates</u>.

1 · 4 Irrational Numbers

1. An irrational number is one whose decimal representation neither <u>repeats</u> nor <u>terminates</u>.

2. Given: $\{-.3, \pi, -4, 0/2, \sqrt{4}, 1/3, 0, 5/0, -\sqrt{13}, 1.52\}$

(a) Which of the above are natural numbers? $\sqrt{4}$ (because $\sqrt{4} = 2$)

(b) Which of the above are integers? $-4, \frac{0}{2}, \sqrt{4}, 0$

11. Convert to decimals:

(a) 33 $-2/9$ $-.2\bar{2}$

(b) $5/6$ $.83\bar{3}$

12. Convert to fractions in lowest terms:

(a) 39 3.875

$3\frac{875}{1000}$

$3\frac{35}{40}$

$3\frac{7}{8}$ or $\frac{31}{8}$

(b) .003

$\frac{3}{1000}$

13. Perform the indicated operations:

(a) 45 (-13.356) ÷ (-6.3)

2.12

(b) 26.3 - 41.8 + 7.005

-8.495

(c) (-.0021)(6.01) -.012621

14. A rational number is one whose decimal representation either repeats or terminates.

1·4 Irrational Numbers

1. An irrational number is one whose decimal representation neither repeats nor terminates.

2. Given: {-.3, π , -4 , 0/2 , $\sqrt{4}$, 1/3 , 0 , 5/0 , $-\sqrt{13}$, 1.52}

   (a) Which of the above are natural numbers? $\sqrt{4} = 2$

   (b) Which of the above are integers? $-4, \sqrt{4}, 0, \frac{0}{2}$

(c) Which of the above are rational numbers? $-.3, -4, \frac{0}{2}, \sqrt{4}, \frac{1}{3}, 0, 1.52$

(d) Which of the above are irrational numbers? $\pi, -\sqrt{13}$

3. Find the length of the diagonal of a rectangle having a length of 8 inches and a width of 5 inches.

8 in.
?
5 in.

$a^2 + b^2 = c^2$
$5^2 + 8^2 = c^2$
$25 + 64 = c^2$
$89 = c^2$
$c = \sqrt{89} \doteq$ 9.43 inches

4. Find the circumference of a circle having a radius of 3.5 cm. (Use $\pi \doteq 3.14$)

3.5 cm

$C = 2\pi r$
$\doteq 2 \times 3.14 \times 3.5$
$\doteq$ 21.98 cm

5. Find the area of a circle having a radius of 2 3/4 inches. (Use $\pi \doteq 22/7$)

$2\frac{3}{4}$ in.

$A = \pi r^2$
$\doteq \frac{22}{7} \cdot \left(\frac{11}{4}\right)^2 \doteq \frac{22}{7} \cdot \frac{121}{16} \doteq \frac{2 \cdot 11 \cdot 11 \cdot 11}{7 \cdot 2 \cdot 2 \cdot 2 \cdot 2} \doteq \frac{1331}{56}$
$\doteq 23\frac{43}{56}$ sq. in.

1 · 5 Real Numbers and Their Properties

1. The set of Real numbers is formed by the union of the set of Rational numbers and the set of Irrational numbers.

(c) Which of the above are rational numbers? $-.3, -4, \frac{0}{2}, \sqrt{4}, \frac{1}{3}, 0, 1.52$

(d) Which of the above are irrational numbers? $\pi, -\sqrt{13}$

3. Find the length of the diagonal of a rectangle having a length of 8 inches and a width of 5 inches.

$A^2 + b^2 = c^2$

$8^2 + 5^2 = c^2$

$64 + 25 = 89$

$c = \sqrt{89}$

9.4339811

4. Find the circumference of a circle having a radius of 3.5 cm.

(Use $\pi \doteq 3.14$)

$C = 2\pi r$

$C = 2(3.14)(3.5)$

$C = 21.98$ cm

5. Find the area of a circle having a radius of 2 3/4 inches.

(Use $\pi \doteq 22/7$)

$A = \pi r^2$

$A = \frac{22}{7}\left(2\frac{3}{4}\right)^2$

$A = \frac{22}{7}\left(\frac{11}{4}\right)^2$

$A = \frac{22}{7} \cdot \frac{121}{16} = \frac{11 \cdot 2 \cdot 11 \cdot 11}{7 \cdot 2 \cdot 2 \cdot 2 \cdot 2} = \frac{1331}{56}$

$A = \frac{1331}{56}$

$A = 23\frac{43}{56}$ sq in

1·5 Real Numbers and Their Properties

1. The set of Real numbers is formed by the union of the set of Rational #'s and the set of Irrational #'s.

2. Locate on the number line: $\{-2, -\sqrt{3}, \sqrt{2}, 5/8, -1.3\}$

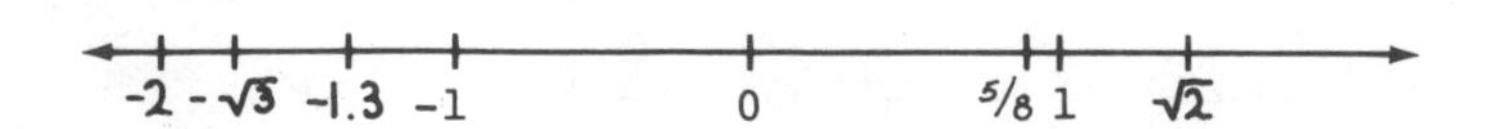

3. Place the correct inequality symbol, < or >, in the blank.

(a) [33] π __>__ 3

(b) 3.3 __<__ $3.\overline{3}$

(c) 7/4 __>__ 1.7

(d) $\sqrt{11}$ __<__ 3.5

4. State the property illustrated by each statement.

symmetric prop. of equality (a) If 1/2 = .5, then .5 = 1/2

reflexive prop. of equality (b) $\sqrt{4} = 2$

substitution (c) If $a = 5$, then $a + 11 = 5 + 11$

transitive prop. of equality (d) If $6 = 12/2$ and $12/2 = \sqrt{36}$, then $6 = \sqrt{36}$

substitution (e) [43] If $x = 3$ and $y = x^2$, then $y = 3^2$

transitive prop. of inequality (f) [41] If $-8 < 0$ and $0 < 4$, then $-8 < 4$

trichotomy (g) If $z \not> 3$, then $z \le 3$

commutative for addition (h) $(6 + 2) + 1 = 1 + (6 + 2)$

associative for multiplication (i) $3.4 \cdot (2.1 \cdot 5.6) = (3.4 \cdot 2.1) \cdot 5.6$

identity for addition (j) [49] $2x + 0 = 2x$

inverse for multiplication (k) $(-2/3)(-3/2) = 1$

distributive property (l) $(-1/2)(5/3) + (-1/2)(3/7)$

$= (-1/2)(5/3 + 3/7)$

5. The additive inverse of a number is its __negative__ or __opposite__.

6. The multiplicative inverse of a nonzero number is its __reciprocal__.

2. Locate on the number line: $\{-2, -\sqrt{3}, \sqrt{2}, 5/8, -1.3\}$

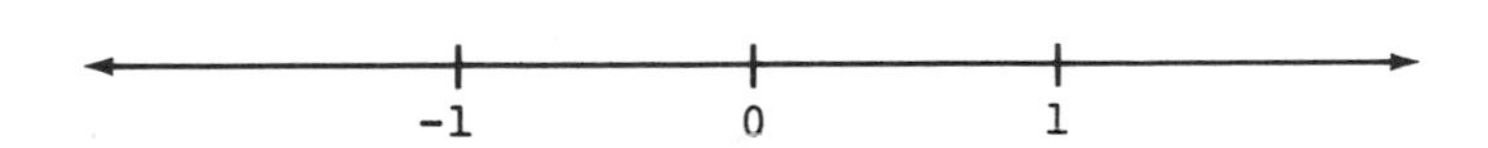

3. Place the correct inequality symbol, < or >, in the blank.

(a) 33 π ___>___ 3 (b) 3.3 ___<___ $3.\overline{3}$

(c) 7/4 ___>___ 1.7 (d) $\sqrt{11}$ ______ 3.5

4. State the property illustrated by each statement.

______________ (a) If $1/2 = .5$, then $.5 = 1/2$

______________ (b) $\sqrt{4} = 2$

______________ (c) If $a = 5$, then $a + 11 = 5 + 11$

______________ (d) If $6 = 12/2$ and $12/2 = \sqrt{36}$, then $6 = \sqrt{36}$

______________ (e) 43 If $x = 3$ and $y = x^2$, then $y = 3^2$

______________ (f) 41 If $-8 < 0$ and $0 < 4$, then $-8 < 4$

______________ (g) If $z \ngtr 3$, then $z \leq 3$

______________ (h) $(6 + 2) + 1 = 1 + (6 + 2)$

______________ (i) $3.4 \cdot (2.1 \cdot 5.6) = (3.4 \cdot 2.1) \cdot 5.6$

______________ (j) 49 $2x + 0 = 2x$

______________ (k) $(-2/3)(-3/2) = 1$

______________ (l) $(-1/2)(5/3) + (-1/2)(3/7)$

$= (-1/2)(5/3 + 3/7)$

5. The additive inverse of a number is its ______________ or ___________.

6. The multiplicative inverse of a nonzero number is its ______________.

1 · 6 Absolute Value, Order of Operations

1. Translate "{ x | x < 3 }" into words.

   The set of all real numbers x such that x is less than three.

2. Graph the set described in problem 1 above.

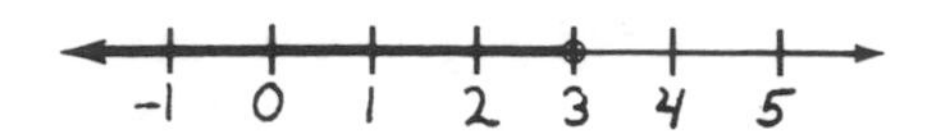

3. Place the appropriate symbol, <, >, or =, in the blank.

   (a) $|a|$ __=__ a, if $a \geq 0$ (b) $|-5|$ __>__ 2

   (c) [55] $|-2|$ __>__ $|0|$ (d) $|\pi|$ __<__ $|-25|$

4. Perform the indicated operations following the rules that govern the order of operations.

   (a) [9] $1.8 \div (1.5)^2 + 0.2$

   $(1.8 \div 2.25) + 0.2 =$
   $.8 + .2 = \boxed{1}$

   (b) $\sqrt{25} \cdot 3.12 - 5.1 \cdot 3.01$

   $(5 \cdot 3.12) - (5.1 \cdot 3.01) =$
   $15.6 - 15.351 = \boxed{.249}$

   (c) $6 - 3(5 \cdot 0 - 8) + 19$

   $6 - 3(-8) + 19 =$
   $6 + 24 + 19 = \boxed{49}$

   (d) $2/3 + 5/8 - 1/4 \cdot 2/5$

   $\frac{2}{3} + \frac{5}{8} - \left(\frac{1}{4} \cdot \frac{2}{5}\right) = \frac{2}{3} + \frac{5}{8} - \frac{1}{10}$

   $= \frac{2}{3} \cdot \frac{40}{40} + \frac{5}{8} \cdot \frac{15}{15} - \frac{1}{10} \cdot \frac{12}{12}$

   $= \frac{80}{120} + \frac{75}{120} - \frac{12}{120} = \frac{143}{120}$ or $\boxed{1 \frac{23}{120}}$

   (e) [21] $[2.1 + 5(0.7)] \div [(0.2)^2 + 0.1]$

   $[2.1 + 3.5] \div [.04 + 0.1] =$
   $5.6 \div .14 = \boxed{40}$

5. Evaluate the following for the given values of the variables.

   (a) $3x + 2y$, for $x = 5$ and $y = -3$

   $3 \cdot 5 + 2 \cdot (-3) =$
   $15 + (-6) = \boxed{9}$

1·6 Absolute Value, Order of Operations

1. Translate "$\{ x \mid x < 3 \}$" into words.

2. Graph the set described in problem 1 above.

3. Place the appropriate symbol, <, >, or =, in the blank.

(a) $|a|$ _____ a, if $a \geq 0$

(b) $|-5|$ _____ 2

(c) [55] $|-2|$ _____ $|0|$

(d) $|\pi|$ _____ $|-25|$

4. Perform the indicated operations following the rules that govern the order of operations.

(a) [9] $1.8 \div (1.5)^2 + 0.2$

(b) $\sqrt{25} \cdot 3.12 - 5.1 \cdot 3.01$

(c) $6 - 3(5 \cdot 0 - 8) + 19$

(d) $2/3 + 5/8 - 1/4 \cdot 2/5$

(e) [21] $[2.1 + 5(0.7)] \div [(0.2)^2 + 0.1]$

5. Evaluate the following for the given values of the variables.

(a) $3x + 2y$, for $x = 5$ and $y = -3$

(b) 59 $-2(x+3)^2-4$, for $x = 1$

$-2(1+3)^2-4=$

$-2(4)^2-4=$

$-2(16)-4=$

$-32-4=\boxed{-36}$

(c) $(-3/4)(2x-5)^2+1$, for $x = (-1/4)$

$\left(\frac{-3}{4}\right)\left[2\cdot\left(\frac{-1}{4}\right)-5\right]^2+1=$

$\left(\frac{-3}{4}\right)\left[-\frac{1}{2}-5\right]^2+1=\left(\frac{-3}{4}\right)\left(-5\frac{1}{2}\right)^2+1=\left(\frac{-3}{4}\right)\left(\frac{-11}{2}\right)^2+1=$

$\left(\frac{-3}{4}\right)\left(\frac{121}{4}\right)+1=-\frac{363}{16}+\frac{16}{16}=\frac{-347}{16}=\boxed{-21\frac{11}{16}}$

(d) $5-\sqrt{x}+3x^2$, for $x = 9$

$5-\sqrt{9}+3\cdot 9^2=$

$5-3+3\cdot 81=$

$5-3+243=\boxed{245}$

(e) 73 $(10x^2+11xy-6y^2)/(5x-2y)$, for $x = 1/2$, $y = 2$

$$\frac{\left[10\left(\frac{1}{2}\right)^2+11\left(\frac{1}{2}\right)(2)-6(2)^2\right]}{\left[5\left(\frac{1}{2}\right)-2(2)\right]}=\frac{10\cdot\frac{1}{4}+11\cdot 1-6\cdot 4}{\frac{5}{2}-4}=$$

$$\frac{\frac{5}{2}+11-24}{\frac{5}{2}-\frac{8}{2}}=\frac{\frac{5}{2}-13}{\frac{-3}{2}}=\frac{\frac{5}{2}-\frac{26}{2}}{\frac{-3}{2}}=\frac{\frac{-21}{2}}{\frac{-3}{2}}=\boxed{7}$$

(b)
59

$-2(x + 3)^2 - 4$, for $x = 1$

(c) $(-3/4)(2x - 5)^2 + 1$, for $x = (-1/4)$

(d) $5 - \sqrt{x} + 3x^2$, for $x = 9$

(e)
73

$(10x^2 + 11xy - 6y^2) / (5x - 2y)$, for $x = 1/2$, $y = 2$

$$\frac{10x^2 + 11xy - 6y^2}{5x - 2y} = \frac{10(\frac{1}{2})^2 + 11(\frac{1}{2})(2) - 6(2)^2}{5(\frac{1}{2}) - 2(2)^2}$$

$$\frac{10 \cdot \frac{1}{4} + 11 \cdot 1 - 6 \cdot 4}{5 \cdot \frac{1}{2} - 4} = \frac{\frac{10}{4} + 11 - 24}{\frac{5}{2} - 4}$$

$$\frac{\frac{5}{2} + 11 - 24}{\frac{5}{2} - \frac{8}{2}} = \frac{\frac{5}{2} - 13}{-\frac{3}{2}} = \frac{\frac{5}{2} - \frac{26}{2}}{-\frac{3}{2}}$$

$$\frac{-\frac{21}{2}}{-\frac{3}{2}}\left(\frac{2}{2}\right) = \frac{-21}{-3} = 7$$

Chapter 1 Self-Test

Consider the sets $A = \{2, 4, 6, 8, 10\}$, $B = \{1, 3, 5, 7, 9\}$, and $C = \{1, 2, 3, 4, 5, \ldots\}$. Answer "True" or "False" to the following statements.

________ 1. $B \subseteq C$

________ 2. $\{8, 10\} \in A$

________ 3. $\emptyset \subseteq A$

________ 4. $A \cap B = \{1, 2, 3, 4, 5, 6, 7, 8, 9, 10\}$

________ 5. $A \cup C = C$

________ 6. C is a finite set.

________ 7. $(A \cup B) \cap C$ is an infinite set.

________ 8. A and B are equal sets.

Perform the indicated operations.

9. $(-3)^2$

10. $3 - (-2) + 5 - 8$

10. $-8/0$

12. -3^2

13. $(-2)(5)(1/2)$

14. $(3/5) \div (-12/25)$

15. $(1.3)(2.25)$

16. $1.6 \div (-2.1)^2 + 0.8$

17. $|5 - (2 - 8 + 1) - 16|$

18. $[(2/3)^4 \div (4/9)] - 1/3$

19. $\sqrt{1/25} \cdot \sqrt{9/49}$

20. $13/15 + 1/9 - 7/24$

Evaluate the following for the given values of the variables.

21. $-3(2 - x)^2 + 2x$, for $x = -2$

22. $2x^3 - \sqrt{3x} - 15$, for $x = 3$

Given the set $A = \{-.4, \pi, -3, 0/5, \sqrt{9}, 1/3, 5/0, -\sqrt{11}, 2.41\}$

23. Which elements of A are integers? ______________________________

24. Which elements of A are irrational numbers? ____________________

State the property of the set of real numbers illustrated by each of the following statements.

__________________________ 25. $45x + 0 = 45x$

__________________________ 26. If $x + 3 = x^2 + 5x + 2$ and $x^2 + 5x + 2 = 8$, then $x + 3 = 8$.

C H A P T E R

2 Linear Equations and Inequalities

2 · 1 Linear Equations

1. A/an equation is an algebraic way of stating that two quantities are equal.

2. In the algebraic expression $7x + 3y - 2x$, 3 is the numerical factor, or coefficient, of the term $3y$, and the terms $7x$ and $-2x$ are called like terms because the variable factors are both x.

3. (a) List the terms in: $3x + 4(y + 1) + 5y - 7/2$ $3x, 4(y+1), 5y, \frac{-7}{2}$

   (b) What is the coefficient of the variable in the term $7x$? 7

   (c) In $3x + 4y - 8x + 3z$, $3x$ and $-8x$ are like terms.

   (d) $1/2\ x + 2/3\ y + 3/4\ x + 1/3\ y =$ $\frac{5}{4}x + y$

   (e) $3(x + 5) + 2(x - 1) =$ $3x + 15 + 2x - 2 = 5x + 13$

   (f) $6c - 6 - 2c - 5 - 3c + 7 =$ $c - 4$

4. If $2x + 2 = 6$, then $1/2\ (2x + 2) = 1/2\ (6)$ because of the Multiplicative Property of Equality.

CHAPTER

2 Linear Equations and Inequalities

2·1 Linear Equations

1. A/an equation is an algebraic way of stating that two quantities are equal.

2. In the algebraic expression $7x + 3y - 2x$, 3 is the numerical factor, or coefficient, of the term $3y$, and the terms $7x$ and $-2x$ are called like terms because the variable factors are both x.

3. (a) List the terms in: $3x + 4(y + 1) + 5y - 7/2$ $3x, 4(y+1), 5y, -\frac{7}{2}$

   (b) What is the coefficient of the variable in the term $7x$? 7

   (c) In $3x + 4y - 8x + 3z$, $3x$ and $-8x$ are like terms.

   (d) $1/2\ x + 2/3\ y + 3/4\ x + 1/3\ y =$ ________

   (e) $3(x + 5) + 2(x - 1) =$ ________

   (f) $6c - 6 - 2c - 5 - 3c + 7 =$ ________

4. If $2x + 2 = 6$, then $1/2\ (2x + 2) = 1/2\ (6)$ because of the ________ Property of Equality.

5. If $2x + 2 = 6$, then $2x + 2 - 2 = 6 - 2$ because of the Additive Property of Equality.

6. The equation $2x - 4 = 0$ is a linear equation in the variable x because it is in the form $ax + b = 0$, $a \neq 0$.

7. $3x + .7 = x - 1.1$

$3x - x + .7 =$ $x - x - 1.1$ Additive Prop. of Equality

$2x + .7 - .7 =$ $-1.1 - .7$ Additive Prop. of Equality

$2x =$ -1.8

$1/2 \cdot 2x =$ $-1.8 \cdot \frac{1}{2}$ Multiplicative Prop. of Equality

$x =$ $-.9$

Check: $3\ (-.9) + .7 = (-.9) - 1.1$

$-2.7 + .7 = -2$

$-2 = -2$

$-.9$ is a solution of the equation because $-2 = -2$ is a true statement of equality.

8. $3(2x - 1) - 2(x + 4) = 8$

$6x - 3 - 2x - 8 = 8$

$4x - 11 = 8$

$4x - 11 + 11 =$ $8 + 11$ Additive Prop. of Equality

$4x =$ 19

$1/4 \cdot 4x =$ $19 \cdot \frac{1}{4}$ Multiplicative Prop. of Equality

$x =$ 4.75

5. If $2x + 2 = 6$, then $2x + 2 - 2 = 6 - 2$ because of the ____________ Property of Equality.

6. The equation $2x - 4 = 0$ is a linear equation in the variable x because ________________________________.

7. $3x + .7 = x - 1.1$

$3x - x + .7 =$ ______________ Additive Prop. of Equality

$2x + .7 - .7 =$ ______________ Additive Prop. of Equality

$2x =$ ______________

$1/2 \cdot 2x =$ ______________ Multiplicative Prop. of Equality

$x =$ ______________

Check: $3\,\underline{(-.9)} + .7 = \underline{(-.9)} - 1.1$

$-2.7 + .7 = -2$

_____ = _____

$-.9$ is a ____________ of the equation because $\underline{-2 = -2}$ is a true statement of equality.

8. $3(2x - 1) - 2(x + 4) = 8$

$6x - 3 - 2x - 8 = 8$

$4x - 11 = 8$

$4x - 11 + 11 =$ ________ Additive Prop. of Equality

$4x =$ ________

$1/4 \cdot 4x =$ ________ Multiplicative Prop. of Equality

$x =$ ________

Check: $3[2(4.75) - 1] - 2[4.75 + 4] = 8$

$3(9.5-1) - 2(8.75) = 8$

$25.5 - 17.5 = 8$

$8 = 8$

4.75 is a solution of the equation because $8 = 8$ is a true statement of equality.

9. $3x - 7 = 6x - 2 - 3x$

$3x - 7 = 3x - 2$

$3x - 3x - 7 = 3x - 3x - 2$ Additive Prop. of Equality

$-7 = -2$

There is no solution to this equation because $-7 = -2$ is not a true statement of equality.

10. $3x - 7 = 6x - 2 - 3x - 5$

$3x - 7 = 3x - 7$

$3x - 3x - 7 = 3x - 3x - 7$ Additive Prop. of Equality

$-7 = -7$

There is an infinite number of solutions (that is, all real numbers will solve the equation) because $-7 = -7$ is a true statement for all values of x.

Check: $3[2(4.75) - 1] - 2[4.75 + 4] = 8$

$\underline{\hspace{6cm}} = 8$

$\underline{\hspace{6cm}} = 8$

$\underline{\hspace{2cm}} = 8$

4.75 is a <u>solution</u> of the equation because <u>8 = 8</u> is a true statement of equality.

9. $3x - 7 = 6x - 2 - 3x$

$3x - 7 = \underline{\hspace{3cm}}$

$3x - 3x - 7 = \underline{\hspace{3cm}}$ Additive Prop. of Equality

$-7 = \underline{\hspace{3cm}}$

There is <u>no solution</u> to this equation because <u>-7 = -2</u> is not a true statement of equality.

10. $3x - 7 = 6x - 2 - 3x - 5$

$3x - 7 = \underline{\hspace{4cm}}$

$3x - 3x - 7 = \underline{\hspace{4cm}}$ Additive Prop. of Equality

$-7 = \underline{\hspace{4cm}}$

There is an <u>infinite number of solutions</u> (that is, all real numbers will solve the equation) because <u>-7 = -7</u> is a true statement for all values of x.

Find the solutions to the following equations and check your solutions.

11. $\frac{-2(2 - 5x)}{3} = \frac{(x + 8)}{3}$

$$3\left[\frac{-2(2-5x)}{3}\right] = 3\left[\frac{x+8}{3}\right]$$

$$-4 + 10x = x + 8$$

$$-4 + 10x - x = x - x + 8$$

$$-4 + 9x = 8$$

$$-4 + 4 + 9x = 8 + 4$$

$$9x = 12$$

$$\frac{1}{9} \cdot 9x = \frac{1}{9} \cdot 12$$

$$x = \frac{12}{9} = \frac{4}{3}$$

Ck: $\frac{-2(2 - 5 \cdot \frac{4}{3})}{3} \stackrel{?}{=} \frac{(\frac{4}{3} + 8)}{3}$

$$\frac{28}{9} = \frac{28}{9}$$

12. $-6x + 10 = 3 - 5x$

$$-6x + 5x + 10 = 3 - 5x + 5x$$

$$-x + 10 = 3$$

$$-x + 10 - 10 = 3 - 10$$

$$-x = -7$$

$$-x \cdot -1 = -7 \cdot -1$$

$$x = \boxed{7}$$

CK: $-6(7) + 10 = 3 - 5(7)$

$$-42 + 10 = 3 - 35$$

$$-32 = -32$$

13. $\frac{x}{3} - \frac{1}{2} = \frac{(x - 1)}{3} - \frac{1}{6}$

$$6\left[\frac{x}{3} - \frac{1}{2}\right] = 6\left[\frac{(x-1)}{3} - \frac{1}{6}\right]$$

$$\frac{6x}{3} - \frac{6}{2} = \frac{6(x-1)}{3} - \frac{6}{6}$$

$$2x - 3 = 2x - 2 - 1$$

$$2x - 3 = 2x - 3$$

There is an infinite number of solutions.

14. $2(3x - 5) = 5x - 3$

$$6x - 10 = 5x - 3$$

$$6x - 5x - 10 = 5x - 5x - 3$$

$$x - 10 = -3$$

$$x - 10 + 10 = -3 + 10$$

$$x = \boxed{7}$$

CK: $2(3 \cdot 7 - 5) = 5(7) - 3$

$$2(21 - 5) = 35 - 3$$

$$2(16) = 32$$

$$32 = 32$$

Find the solutions to the following equations and check your solutions.

11. $(3)\ \frac{-2(2 - 5x)}{3} = \frac{(x + 8)}{3}\ (3)$

$-4 + 10x = x + 8$

$+4 \quad -x \quad\quad -x + 4$

$9x = 12$

$\frac{9x}{9} = \frac{12}{9}$

$x = \frac{4}{3}$

(cross multiplying also works)

12. $-6x + 10 = 3 - 5x$

$+5x - 10 = -10 + 5x$

$-x = -7$

$x = 7$

13. $\frac{x}{3} - \frac{1}{2} = \frac{(x - 1)}{3} - \frac{1}{6}$

14. $2(3x - 5) = 5x - 3$

15. $-.5x + 3.2 = 1.3x + 4.4$

51

$-.5x + 3.2 - 3.2 = 1.3x + 4.4 - 3.2$

$-.5x = 1.3x + 1.2$

$-.5x - 1.3x = 1.3x - 1.3x + 1.2$

$-1.8x = 1.2$

$-\frac{1}{1.8} \cdot -1.8x = -\frac{1}{1.8} \cdot 1.2$

$x = \frac{-1.2}{1.8}$

$x = \boxed{\frac{-2}{3}}$

16. $3x + 8 = x + 2(x + 3)$

$3x + 8 = x + 2x + 6$

$3x + 8 = 3x + 6$

$3x - 3x + 8 = 3x - 3x + 6$

$8 = 6$

There are no solutions.

17. $.3(x - 7) + .4(2x + 1) = .2(x + 5)$

$.3x - 2.1 + .8x + .4 = .2x + 1$

$1.1x - 1.7 = .2x + 1$

$1.1x - 1.7 + 1.7 = .2x + 1 + 1.7$

$1.1x = .2x + 2.7$

$1.1x - .2x = .2x - .2x + 2.7$

$.9x = 2.7$

$\frac{1}{.9} \cdot .9x = \frac{1}{.9} \cdot 2.7$

$x = \boxed{3}$

18. $-3(x + 2) - (x - 3) = -7$

$-3x - 6 - x + 3 = -7$

$-4x - 3 = -7$

$-4x - 3 + 3 = -7 + 3$

$-4x = -4$

$-\frac{1}{4} \cdot -4x = -\frac{1}{4} \cdot -4$

$x = \boxed{1}$

19. In problems 11 through 18 above, which equations are equivalent?

Equations 12 and 14 are equivalent because they have the same Solution.

2 · 2 Linear Inequalities

1. "$3x + 2 < 5$" is an example of a/an linear inequality.

2. The symbol "<" is read "less than."

15. | 51

$-.5x + 3.2 = 1.3x + 4.4$

$-.5x = 1.3x + 1.2$

$-1.8x = 1.2$

$\frac{-1.8x}{-1.8} = \frac{1.2}{-1.8}$

$x = -\frac{2}{3}$

16. $3x + 8 = x + 2(x + 3)$

$3x + 8 = x + 2x + 6$

$3x + 8 = 3x + 6$

$8 \neq 6$

$\emptyset$

17. $.3(x - 7) + .4(2x + 1) = .2(x + 5)$

$.3x - 2.1 + .8x + .4 = .2x + 1.0$

$1.1x - 1.7 = .2x + 1$

$1.1x = .2x + 1 + 1.7$

$.9x = 2.7$

$\frac{.9x}{.9} = \frac{2.7}{.9}$

$x = 3$

18. $-3(x + 2) - (x - 3) = -7$

$-3x - 6 - x + 3 = -7$

$-4x - 3 = -7$

$-4x = -4$

$\frac{-4x}{-4} = \frac{-4}{-4}$

$x = 1$

19. In problems 11 through 18 above, which equations are equivalent?

2·2 Linear Inequalities

1. "$3x + 2 < 5$" is an example of a/an linear inequality.

2. The symbol "<" is read "less than."

3. "$x \geq 3$" is read: "x is greater than or equal to 3."

Translate the following statements into algebraic expressions which use the inequality symbols $>$, $<$, $\geq$, or $\leq$.

4. Six times a number is greater than 30.5. $6n > 30.5$

5. The sum of x and y is less than or equal to the product of x and y.

   $x + y \leq xy$

6. Negative six is less than six. $-6 < 6$

7. Twice a number is greater than or equal to the number decreased by fifteen. $2n \geq n - 15$

Match the inequality to its corresponding graph.

Answer	Inequality	Graph
d	8. $x < 3$	a. −2 −1 0 1 2 3 4
b	9. $x \geq 3$	b. −2 −1 0 1 2 3 4
e	10. $x \leq 3$	c. −2 −1 0 1 2 3 4
a	11. $x > 3$	d. −2 −1 0 1 2 3 4
c	12. $0 \leq x < 3$	e. −2 −1 0 1 2 3 4

13. $-3 < 7$

-3 (-2) ? 7 (-2) (Multiply both sides by -2)

6 $>$ -14 ($<$, $>$, $\leq$, $\geq$, $=$) Choose the correct symbol.

NOTE: From the above example it can be concluded that: When multiplying both sides of an inequality by a negative number, the inequality symbol must be reversed.

3. "$x \geq 3$" is read: "x is greater than or equal to 3."

Translate the following statements into algebraic expressions which use the inequality symbols $>$, $<$, $\geq$, or $\leq$.

4. Six times a number is greater than 30.5. $6x > 30.5$

5. The sum of x and y is less than or equal to the product of x and y.

   $x + y \leq xy$

6. Negative six is less than six. $-6 < 6$

7. Twice a number is greater than or equal to the number decreased by fifteen. $2x \geq x - 15$

Match the inequality to its corresponding graph.

d 8. $x < 3$

b 9. $x \geq 3$

e 10. $x \leq 3$

a 11. $x > 3$

c 12. $0 \leq x < 3$

a. -2 -1 0 1 2 3 4

b. -2 -1 0 1 2 3 4

c. -2 -1 0 1 2 3 4

d. -2 -1 0 1 2 3 4

e. -2 -1 0 1 2 3 4

13. $-3 < 7$

-3 (-2) ? 7(-2) (Multiply both sides by -2)

6 > -14 ($<$, $>$, $\leq$, $\geq$, $=$) Choose the correct symbol.

NOTE: From the above example it can be concluded that: When multiplying both sides of an inequality by a negative number, the inequality symbol must be reversed.

Find the solutions of the following inequalities and graph the solutions on the number line.

14. $4(x - 2) - (x + 1) < 6$

$4x - 8 - x - 1 < 6$

$3x - 9 + 9 < 6 + 9$

$3x < 15$

$\frac{1}{3} \cdot 3x < \frac{1}{3} \cdot 15$

$\boxed{x < 5}$

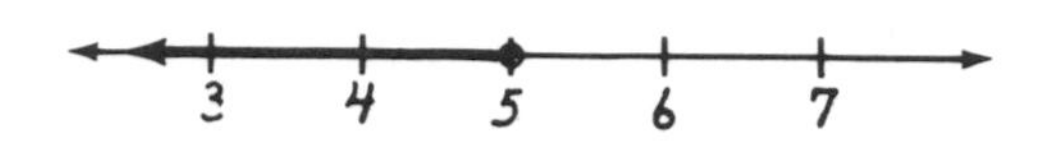

15. $-3x + 5 \geq 23$

$-3x + 5 - 5 \geq 23 - 5$

$-3x \geq 18$

$-\frac{1}{3} \cdot -3x \leq -\frac{1}{3} \cdot 18$

$\boxed{x \leq -6}$

16. / 37 $1/5\ x - 3 \leq 5/4 - 3/2\ x$

$\frac{1}{5}x - 3 + 3 \leq \frac{5}{4} + 3 - \frac{3}{2}x$

$\frac{1}{5}x \leq \frac{17}{4} - \frac{3}{2}x$

$\frac{1}{5}x + \frac{3}{2}x \leq \frac{17}{4} - \frac{3}{2}x + \frac{3}{2}x$

$\frac{10}{17} \cdot \frac{17}{10}x \leq \frac{17}{4} \cdot \frac{10}{17}$

$\boxed{x \leq 2.5}$

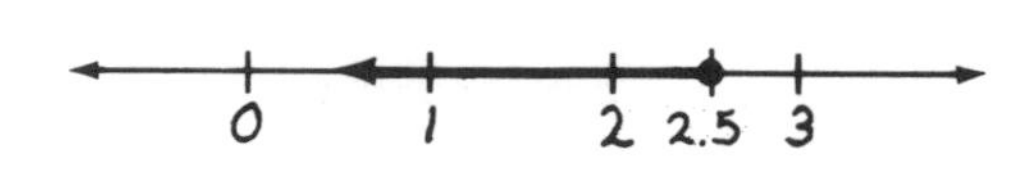

17. / 40 $.19 - .35x \leq .17x + .06$

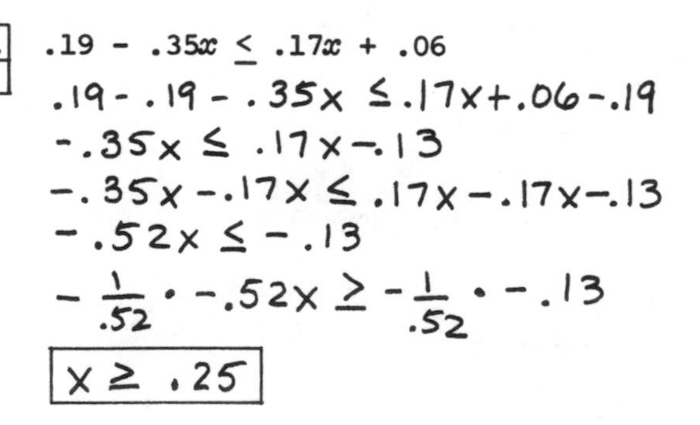

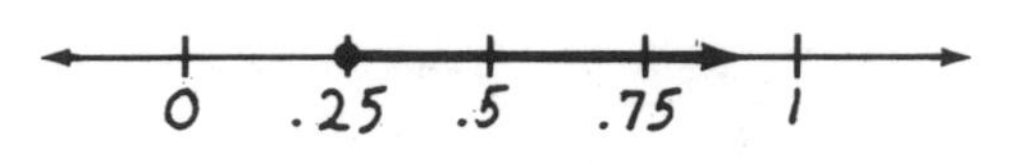

Find the solutions of the following inequalities and graph the solutions on the number line.

14. $4(x - 2) - (x + 1) < 6$

$4x - 8 - x - 1 < 6$

$3x - 9 < 6$

$\frac{3x}{3} < \frac{15}{3}$

$x < 5$

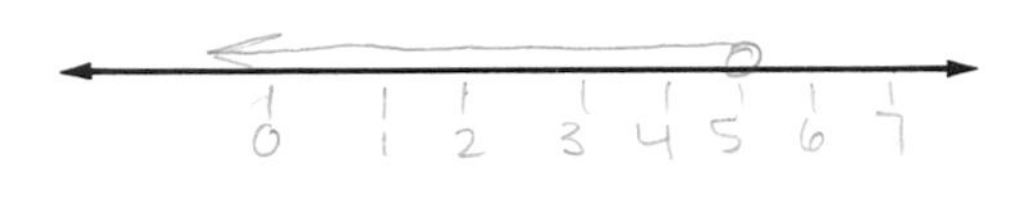

15. $-3x + 5 \geq 23$

$\frac{-3x}{-3} \geq \frac{18}{-3}$

$x \leq -6$

-7 -6 -5 -4 -3 -2 -1

16. / 37 $1/5\ x - 3 \leq 5/4 - 3/2\ x$

$\frac{1}{5}x - 3 \leq \frac{5}{4} - \frac{3}{2}x$

$\frac{1}{5}x \leq \frac{5}{4} - \frac{3}{2}x + 3$

$20(\frac{1}{5}x) \leq 20(\frac{5}{4}) - 20(\frac{3}{2}x) + 20(3)$

$4x \leq 25 - 30x + 60$

$\frac{34x}{34} \leq \frac{85}{34}$

$x \leq 2.5$

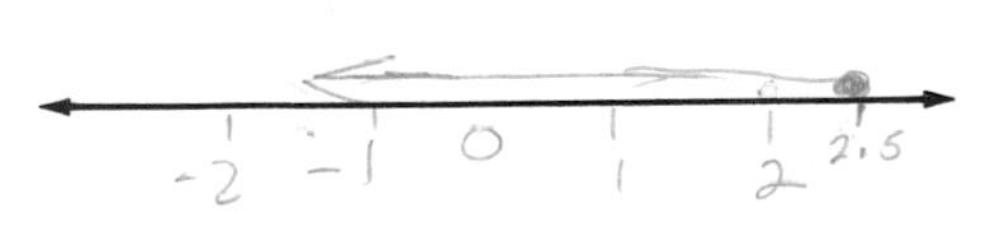

17. / 40 $.19 - .35x \leq .17x + .06$

$-.35x \leq .17x + .06 - .19$

$-.35x - .17x \leq -.13$

$\frac{-.52x}{-.52} \leq \frac{-.13}{-.52}$

$x \geq .25$

18. $7x - 2(5x + 1) > x + 1$

$7x - 10x - 2 > x + 1$

$-3x - 2 + 2 > x + 1 + 2$

$-3x - x > x - x + 3$

$-4x > 3$

$-\frac{1}{4} \cdot -4x < -\frac{1}{4} \cdot 3$

$\boxed{x < -\frac{3}{4}}$

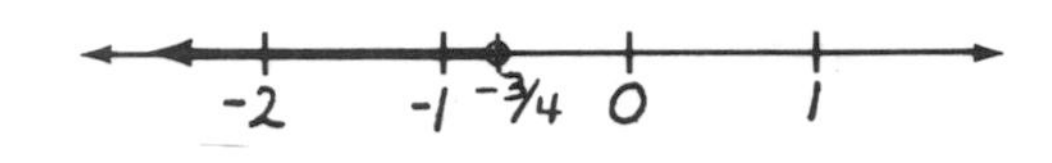

19. $1/2\ (3 - x) \le -4$

$\frac{3}{2} - \frac{1}{2}x \le -4$

$-\frac{3}{2} + \frac{3}{2} - \frac{1}{2}x \le -\frac{3}{2} - 4$

$-\frac{1}{2}x \le -\frac{11}{2}$

$-\frac{2}{1} \cdot -\frac{1}{2}x \ge -\frac{2}{1} \cdot -\frac{11}{2}$

$\boxed{x \ge 11}$

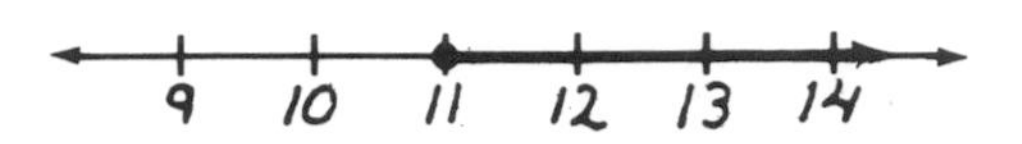

20. $-3 < 1 - 2x \le 7$

$-3 - 1 < 1 - 1 - 2x \le 7 - 1$

$-4 < -2x \le 6$

$-\frac{1}{2} \cdot -4 > -\frac{1}{2} \cdot -2x \ge -\frac{1}{2} \cdot 6$

$\boxed{2 > x \ge -3}$ OR $-3 \le x < 2$

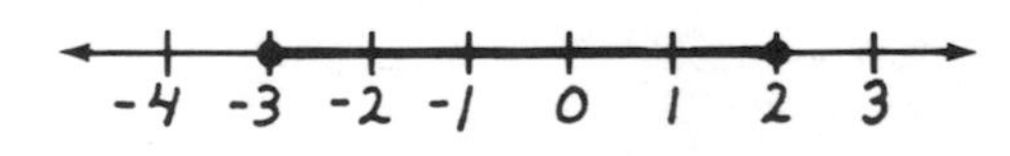

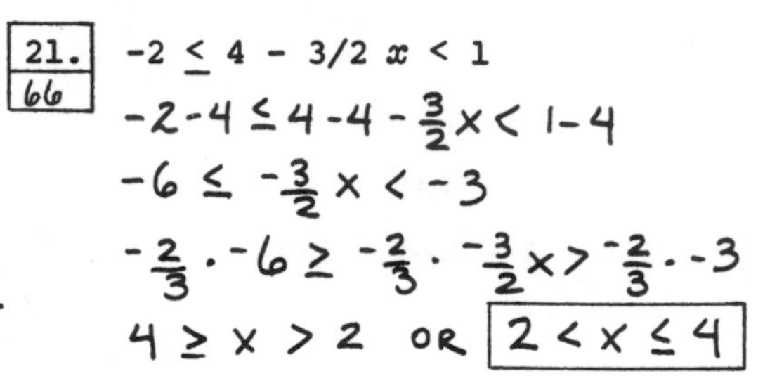

21. (66) $-2 \le 4 - 3/2\ x < 1$

$-2 - 4 \le 4 - 4 - \frac{3}{2}x < 1 - 4$

$-6 \le -\frac{3}{2}x < -3$

$-\frac{2}{3} \cdot -6 \ge -\frac{2}{3} \cdot -\frac{3}{2}x > -\frac{2}{3} \cdot -3$

$4 \ge x > 2$ OR $\boxed{2 < x \le 4}$

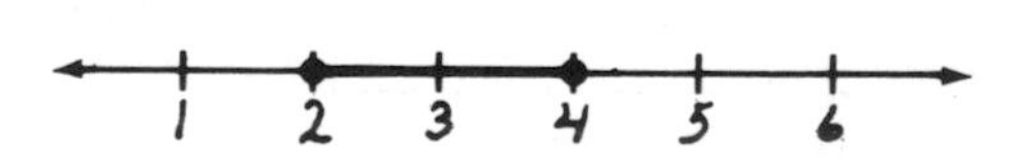

22. When two inequalities are connected with the words "and" or "or," the resulting statement is called a <u>compound inequality</u>.

23. When the word "and" is included in a compound statement, then a/an <u>intersection</u> of sets is implied.

24. When the word "or" is included in a compound statement, then a/an <u>union</u> of sets is implied.

18. $7x - 2(5x + 1) > x + 1$

$7x - 10x - 2 > x + 1$

$-x - 3x > 3$

$-4x > 3$

$\frac{-4x}{-4} \quad \frac{3}{-4}$

$x < -\frac{3}{4}$

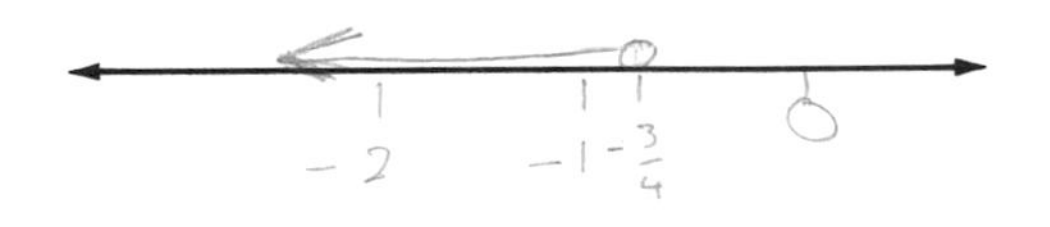

19. $1/2\ (3 - x) \leq -4$

$\left(\frac{1}{2}\right)\frac{3}{1} - \frac{1}{2}x \leq -4$

$\frac{3}{2} - \frac{1}{2}x \leq -4 - \frac{3}{2}$

$(2) - \frac{1}{2}x \leq (2) - 4 - 2\left(\frac{3}{2}\right)$

$-x \leq -8 - 3$

$-x \leq -11$

$x \geq 11$

20. $-3 < 1 - 2x \leq 7$

$\frac{-4}{-2} < \frac{-2x}{-2} \leq \frac{6}{-2}$

$2 > x \geq -3$

21. | 66 $-2 \leq 4 - 3/2\ x < 1$

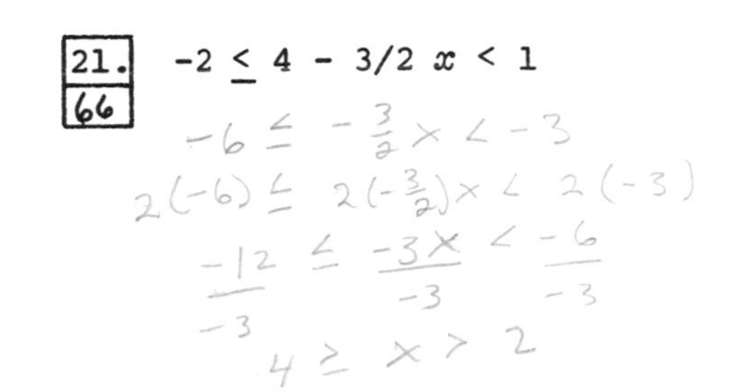

$-6 \leq -\frac{3}{2}x < -3$

$2(-6) \leq 2(-\frac{3}{2})x < 2(-3)$

$\frac{-12}{-3} \leq \frac{-3x}{-3} < \frac{-6}{-3}$

$4 \geq x > 2$

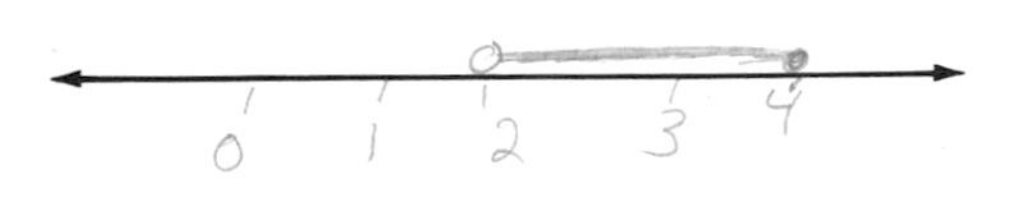

22. When two inequalities are connected with the words "and" or "or," the resulting statement is called a compound inequality.

23. When the word "and" is included in a compound statement, then a/an intersection of sets is implied.

24. When the word "or" is included in a compound statement, then a/an union of sets is implied.

Find the solutions of the following compound inequalities. Express the solution in terms of (a) set builder notation and (b) graph.

25. (67) $3x - 1 < 2$ and $5x + 4 > 9$

$3x < 3$ $5x > 5$

$x < 1$ and $x > 1$

(a) $\{x \mid x < 1 \text{ and } x > 1\}$ $\emptyset$

(b)

26. $3x - 1 < 2$ or $5x + 4 > 9$

$3x < 3$ $5x > 5$

$x < 1$ OR $x > 1$

(a) $\{x \mid x < 1 \text{ OR } x > 1\}$

(b) -1 0 1 2

27. $-7x - 5 \geq 9$ or $4x - 11 \leq 7$

$7x \geq 14$ $4x \leq 18$

$x \leq -2$ OR $x \leq \frac{9}{2}$

* Remember: the union of $x \leq -2$ and $x \leq \frac{9}{2}$ will be $x \leq \frac{9}{2}$.

(a) $\{x \mid x \leq \frac{9}{2}\}$

(b) 4 9/2 5 6

28. $3 \leq 2x < 8$ and $-4 < x - 6 \leq 1$

$\frac{3}{2} \leq x < 4$ and $2 < x \leq 7$

* Remember: "and" implies intersection, or where the two graphs overlap.

(a) $\{x \mid 2 < x < 4\}$

(b)

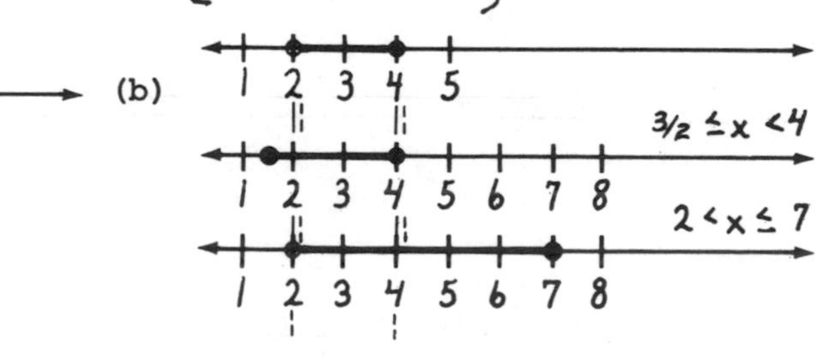

2·3 Literal Equations

1. Equations such as $C = 2\pi r$ and $P = 2L + 2W$ are examples of literal equations.

2. (1) Solve the distance formula $D = RT$ for R.

$D = RT$

$\frac{1}{T} \cdot D = R \cdot \frac{1}{T} \cdot T$

$\boxed{\frac{D}{T} = R}$

Find the solutions of the following compound inequalities. Express the solution in terms of (a) set builder notation and (b) graph.

25. $3x - 1 < 2$ and $5x + 4 > 9$

$\frac{3x}{3} < \frac{3}{3}$ $\frac{5x}{5} > \frac{5}{5}$

$\{x \mid x < 1$ and $x > 1\}$

(a)

(b) ∅

26. $3x - 1 < 2$ or $5x + 4 > 9$

$\frac{3x}{3} < \frac{3}{3}$ $\frac{5x}{5} > \frac{5}{5}$

$\{x \mid x < 1$ or $x > 1\}$

(a)

(b) -1 0 1 2 3

27. $-7x - 5 \geq 9$ or $4x - 11 \leq 7$

$\frac{-7x}{-7} \geq \frac{14}{-7}$ or $\frac{4x}{4} \leq \frac{18}{4}$

$x \leq -2$ or $x \leq \frac{9}{2}$

(a)

(b) 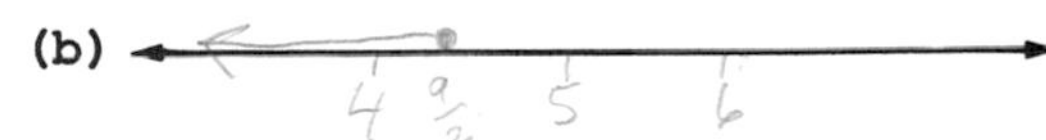

4 $\frac{9}{2}$ 5 6

28. $3 \leq 2x < 8$ and $-4 < x - 6 \leq 1$

$\frac{3}{2} \leq \frac{2x}{2} < \frac{8}{2}$ $2 < x \leq 7$

$\frac{3}{2} \leq x < 4$

(a) $\{x \mid$

(b) 2 3 4

2·3 Literal Equations

1. Equations such as $C = 2\pi r$ and $P = 2L + 2W$ are examples of literal equations.

2. Solve the distance formula $D = RT$ for R.

$\frac{D}{T} = \frac{RT}{T}$

$\frac{D}{T} = R$

3. | 11

Solve $P = 2L + 2W$ for L.

$P - 2W = 2L + 2W - 2W$

$\frac{1}{2}(P - 2W) = (2L)\frac{1}{2}$

$\boxed{\frac{1}{2}(P - 2W) = L}$

4. | 15

Solve $y = mx + b$ for m.

$y - b = mx + b - b$

$\frac{1}{x}(y - b) = \frac{1}{x}(mx)$

$\boxed{\frac{1}{x}(y - b) = m}$

5. Solve $A = 1/2\, h(b_1 + b_2)$ for h.

$A = \frac{1}{2} h (b_1 + b_2)$

$2A = 2 \cdot \frac{1}{2} h (b_1 + b_2)$

$\frac{1}{b_1 + b_2} \cdot 2A = h(b_1 + b_2) \cdot \frac{1}{b_1 + b_2}$

$\boxed{\frac{2A}{b_1 + b_2} = h}$

6. (a) Using the formula $I = PRT$, find I if $P = \$1500$, $R = 12\%$, and $T = 3$ months.

$I = 1500 \cdot .12 \cdot \frac{3}{12}$

$\boxed{I = \$45}$

(b) When using the formula $I = PRT$, T is always expressed in terms of what unit? years

7. Using the formula $V = 1/3 \cdot \pi r^2 h$, find h if $r = 9$ in. and $V = 150$ cu in. (Use $\pi = 3.14$)

$150 = \frac{1}{3}(3.14)(9^2)h$

$3 \cdot 150 = 3 \cdot \frac{1}{3}(3.14)(81)h$

$450 = 254.34h$

$\frac{1}{254.34} \cdot 450 = \frac{1}{254.34} h$

$\boxed{1.769 \text{ in.} = h}$

3. / 11 Solve $P = 2L + 2W$ for L.

$$\frac{P - 2W}{2} = \frac{2L}{2}$$

$$\frac{P - 2W}{2} = L$$

4. / 15 Solve $y = mx + b$ for m.

$$\frac{y - b}{x} = \frac{mx}{x}$$

$$\frac{y - b}{x} = m$$

5. Solve $A = 1/2\ h(b_1 + b_2)$ for h.

$$2(A) = 2\left(\frac{1}{2}\right)h(b_1 + b_2)$$

$$\frac{2A}{b_1 + b_2} = \frac{h(b_1 + b_2)}{b_1 + b_2}$$

$$\frac{2A}{b_1 + b_2} = h$$

6. (a) Using the formula $I = PRT$, find I if $P = \$1500$, $R = 12\%$, and $T = 3$ months.

$$I = (1500).12\left(\frac{3}{12}\right)$$

$$I = \frac{540}{12}$$

$$I = 845$$

(b) When using the formula $I = PRT$, T is always expressed in terms of what unit? years

7. Using the formula $V = 1/3 \cdot \pi r^2 h$, find h if $r = 9$ in. and $V = 150$ cu in. (Use $\pi = 3.14$)

8. | 32 Jack Rabbit deposits \$2400 in a savings account paying 8% simple interest. The amount of interest paid out was \$1152. How long did Jack leave his money in the savings account?

$I = \$1152$
$P = \$2400$
$R = 8\%$
$T = ?$

$I = P \cdot R \cdot T$
$1152 = 2400 \cdot .08 \cdot T$
$1152 = 192T$
$\frac{1}{192} \cdot 1152 = \frac{1}{192} \cdot 192T$
$6 = T$
6 years

9. A lake has a temperature of 5 degrees Celsius. What is the equivalent temperature if measured in degrees Fahrenheit?

$F = \frac{9}{5}C + 32$
$F = \frac{9}{5} \cdot 5 + 32$
$F = 9 + 32 =$ **41° Fahrenheit**

2·4 Applications

1. Match each phrase with its algebraic equivalent.

	Phrase		Algebraic Equivalent
d	5 less than x	(a)	$5 - 3x$
a	5 minus three times x	(b)	$5 - x$
i	three times x decreased by 5	(c)	$3x + 4$
g	8 subtracted from x	(d)	$x - 5$
j	the difference of twice x and 6	(e)	$5x$
b	5 less x	(f)	$5/x$
e	the product of x and 5	(g)	$x - 8$

8. 32 Jack Rabbit deposits \$2400 in a savings account paying 8% simple interest. The amount of interest paid out was \$1152. How long did Jack leave his money in the savings account?

$1152 = 2400(.08)T$

$\frac{1152}{192} = \frac{192T}{192}$

$6 = T$

6 years

9. A lake has a temperature of 5 degrees Celsius. What is the equivalent temperature if measured in degrees Fahrenheit?

$F = \frac{9}{5}C + 32$

$F = 5\left(\frac{9}{5}\right) + 32$

$F = 9 + 32$

$F = 41°$

2·4 Applications

1. Match each phrase with its algebraic equivalent.

	Phrase	Algebraic Equivalent
d	5 less than x	(a) $5 - 3x$
a	5 minus three times x	(b) $5 - x$
i	three times x decreased by 5	(c) $3x + 4$
g	8 subtracted from x	(d) $x - 5$
j	the difference of twice x and 6	(e) $5x$
b	5 less x	(f) $5/x$
e	the product of x and 5	(g) $x - 8$

h 3 more than 5 times x — (h) $3 + 5x$

c the sum of three times x and 4 — (i) $3x - 5$

f the quotient of 5 and x — (j) $2x - 6$

2. (9) The sum of two numbers is 5. The larger number is 17 more than twice the smaller number. What are the numbers?

x = smaller number
$2x+17$ = larger number

$$x + 2x + 17 = 5$$
$$3x + 17 = 5$$
$$3x = -12$$
$$x = -4$$

$x = \boxed{-4}$ = smaller number

$2x+17 = \boxed{9}$ = larger number

3. (17) Find three consecutive integers with the property that the second integer plus the third integer is 18 more than the first integer.

x = 1st integer
$x+1$ = 2nd integer
$x+2$ = 3rd integer

$$(x+1) + (x+2) = 18 + x$$
$$2x + 3 = 18 + x$$
$x = \boxed{15}$
$x+1 = \boxed{16}$
$x+2 = \boxed{17}$

4. (23) The perimeter of a triangle is 17 centimeters. The second side is twice the length of the first side, and the third side is one centimeter less than 3 times the length of the first side. What are the lengths of the sides of the triangle?

x = length of 1st side
$2x$ = length of 2nd side
$3x-1$ = length of 3rd side

$$(x) + (2x) + (3x-1) = 17$$
$$6x - 1 = 17$$
$$x = 3$$

$x = \boxed{3\text{ cm}}$ = length 1st
$2x = \boxed{6\text{ cm}}$ = length 2nd
$3x-1 = \boxed{8\text{ cm}}$ = length 3rd

h 3 more than 5 times x (h) $3 + 5x$

c the sum of three times x and 4 (i) $3x - 5$

f the quotient of 5 and x (j) $2x - 6$

2. 9 The sum of two numbers is 5. The larger number is 17 more than twice the smaller number. What are the numbers?

$5 = (17+2x) + x$

$5 = 3x + 17$

$\frac{-12}{3} = \frac{3x}{3}$ $-4 = x$

$2x + 17 = 9$

3. 17 Find three consecutive integers with the property that the second integer plus the third integer is 18 more than the first integer.

$(x+1)+(x+2) = 18 + x$

$2x + 3 = 18 + x$

$2x = 15 + x$

$x = 15$ 15, 16, 17

$16 + 17 = 33$

$33 - 18 = 15$

4. 23 The perimeter of a triangle is 17 centimeters. The second side is twice the length of the first side, and the third side is one centimeter less than 3 times the length of the first side. What are the lengths of the sides of the triangle?

P = 17

2L, 3L − 1, L

3, 6, 8 = 17

$2L + L + 3L - 1 = 17$

$6L - 1 = 17$

$\frac{6L}{6} = \frac{18}{6}$

$L = 3$

5. 27 Ramon has \$4.80 in quarters and dimes. He has a total of 30 coins. How many quarters and how many dimes does he have?

x = number of quarters
$30-x$ = number of dimes
$.25x$ = value of quarters
$.10(30-x)$ = value of dimes

$$.25x + .10(30-x) = 4.80$$
$$.25x + 3 - .10x = 4.80$$
$$.15x = 1.80$$
$$x = \boxed{12 \text{ quarters}}$$
$$30-x = \boxed{18 \text{ dimes}}$$

6. 35 Pat is six years older than his friend Gary. Four years ago Gary's age was 7 years more than half Pat's age. What are their present ages?

x = Gary's age now
$x+6$ = Pat's age now
$x-4$ = Gary 4 yrs. ago
$(x+6)-4$ = Pat 4 yrs. ago

$$x-4 = 7 + \frac{1}{2}(x+6-4)$$
$$x-4 = 7 + \frac{1}{2}x + 1$$
$$x-4 = 8 + \frac{1}{2}x$$
$$\frac{1}{2}x = 12$$
$$x = \boxed{24 = \text{Gary's age now}}$$
$$x+6 = \boxed{30 = \text{Pat's age now}}$$

7. 37 Henry leaves town at noon on a bicycle traveling west at a rate of 15 mph. At 4:00 p.m. Sarah leaves town in a car traveling in the same direction at a rate of 45 mph. How many hours will it take Sarah to catch Henry?

x = Sarah's time
$x+4$ = Henry's time (since he left 4 hours earlier).

$d = r \cdot t$
Henry's distance $= 15(x+4)$
Sarah's distance $= 45(x)$

$$15(x+4) = 45x$$
$$15x + 60 = 45x$$
$$60 = 30x$$
$$\boxed{x = 2 \text{ hrs}}$$

5. 27 Ramon has $4.80 in quarters and dimes. He has a total of 30 coins. How many quarters and how many dimes does he have?

$x = Q$

$30 - x = D$

$.25x = Q$

$.10(30 - x) = D$

$.25x + .10(30 - x) = 4.80$

$.25x + 3.0 - .10x = 4.80$

$.15x = 1.80$

$\frac{.15x}{.15} = \frac{1.80}{.15}$

$x = 12$ quarters

18 dimes

6. 35 Pat is six years older than his friend Gary. Four years ago Gary's age was 7 years more than half Pat's age. What are their present ages?

X = Gary now = 24

$X + 6$ = Pat now = 30

$X - 4$ = Gary } 4 yrs ago = 20

$X + 6 - 4$ = Pat } 4 yrs ago = 26

$x - 4 = 7 + \frac{1}{2}(x + 6 - 4)$

$x - 4 = 7 + \frac{1}{2}x + 3 - 2$

$x - 4 = 7 + \frac{1}{2}x + 1$

$x = 4 + 7 + 1 + \frac{1}{2}x$

$x = 12 + \frac{1}{2}x$

$2x = 24 + x$

$x = 24$

7. 37 Henry leaves town at noon on a bicycle traveling west at a rate of 15 mph. At 4:00 p.m. Sarah leaves town in a car traveling in the same direction at a rate of 45 mph. How many hours will it take Sarah to catch Henry?

	D	r	t
Henry	15(t+4)	15	t+4
Sarah	45t	45	t

$15t + 60 = 45t$

$60 = 45t - 15t$

$\frac{60}{30} = \frac{30t}{30}$

2 hrs $= t$

2·5 Absolute Value Equations

1. (a) MOST absolute value equations will have two solutions.

   (b) The only time that an expression of the form $|ax + b| = c$ has only one solution is when c is equal to zero.

   (c) The solution of $|x| = -5$ is ϕ. because the absolute value of a number is always positive.

2. Find the solutions of the following equation.

$3|2x + 1| - 5 = 7$

$3|2x + 1| - 5 + 5 = 7+5$

$1/3 \cdot 3|2x + 1| = \frac{1}{3} \cdot 12$

$|2x + 1| = 4$

$2x + 1 = 4$	or	$2x + 1 = -4$
$2x + 1 - 1 = 4-1$	or	$2x + 1 - 1 = -4-1$
$2x = 3$	or	$2x = -5$
$1/2 \cdot 2x = \frac{1}{2} \cdot 3$	or	$1/2 \cdot 2x = \frac{1}{2} \cdot -5$
$x = \frac{3}{2}$	or	$x = -\frac{5}{2}$

Check results:

$3|2\left(\frac{3}{2}\right) + 1| - 5 = 7$	$3|2\left(-\frac{5}{2}\right) + 1| - 5 = 7$
$3(4)-5 = 7$	$3(+4)-5 = 7$
$12-5 = 7$	$12-5 = 7$
$7 = 7$	$7 = 7$

2·5 Absolute Value Equations

1. (a) MOST absolute value equations will have ___Two___ solutions.

   (b) The only time that an expression of the form $|ax + b| = c$ has only one solution is when c is equal to ___0___.

   (c) The solution of $|x| = -5$ is ___$\emptyset$___.

2. Find the solutions of the following equation.

$3|2x + 1| - 5 = 7$

$3|2x + 1| - 5 + 5 =$ ___7+5___

$1/3 \cdot 3|2x + 1| =$ ___$\frac{1}{3} \cdot 12$___

$|2x + 1| =$ ___4___

$2x + 1 =$ ___4___	or	$2x + 1 =$ ___-4___
$2x + 1 - 1 =$ ___4-1___	or	$2x + 1 - 1 =$ ___-4-1___
$2x =$ ___3___	or	$2x =$ ___-5___
$1/2 \cdot 2x =$ ___$\frac{3}{2}$___	or	$1/2 \cdot 2x =$ ___$-\frac{5}{2}$___
$x =$ ___$\frac{3}{2}$___	or	$x =$ ___$-\frac{5}{2}$___

Check results:

$3|2$ ___$(\frac{3}{2})$___ $+ 1| - 5 = 7$	$3|2$ ___$(-\frac{5}{2})$___ $+ 1| - 5 = 7$
___3(4)+5___ $= 7$	___3(+4)-5___ $= 7$
___12-5___ $= 7$	___12-5___ $= 7$
$7 = 7$	$7 = 7$

$\frac{2}{1} - \frac{5}{2} = -\frac{10}{2} = -5+1 = -4$

Find the solutions of the following equations.

3. | 21

$4|-s| = 24$

$\frac{1}{4} \cdot 4|-s| = \frac{1}{4} \cdot 24$

$|-s| = 6$

$-s = 6$ OR $-s = -6$

$-1 \cdot -s = -1 \cdot 6$ $\quad -1 \cdot -s = -1 \cdot -6$

$s = -6$ OR $s = 6$

4. | 41

$-3|x + 4| = -21$

$-\frac{1}{3} \cdot -3|x+4| = -\frac{1}{3} \cdot -21$

$|x+4| = 7$

$x+4 = 7$ OR $x+4 = -7$

$x+4-4 = 7-4$ $\quad x+4-4 = -7-4$

$x = 3$ OR $x = -11$

5. $8 - 3|3x - 1| = -13$

$8-8-3|3x-1| = -13-8$

$-3|3x-1| = -21$

$-\frac{1}{3} \cdot -3|3x-1| = -\frac{1}{3} \cdot -21$

$|3x-1| = 7$

$3x-1 = 7$ OR $3x-1 = -7$

$3x-1+1 = 7+1$ $\quad 3x-1+1 = -7+1$

$3x = 8$ $\quad 3x = -6$

$\frac{1}{3} \cdot 3x = \frac{1}{3} \cdot 8$ $\quad \frac{1}{3} \cdot 3x = -6 \cdot \frac{1}{3}$

$x = \frac{8}{3}$ OR $x = -2$

6. | 69

$|1/2\, x + 5| = |x - 2|$

$\frac{1}{2}x + 5 = x - 2$ OR $\frac{1}{2}x + 5 = -(x-2)$

$\frac{1}{2}x - \frac{1}{2}x + 5 = x - \frac{1}{2}x - 2$ $\quad \frac{1}{2}x + 5 = -x + 2$

$5 = \frac{1}{2}x - 2$ $\quad \frac{1}{2}x + x + 5 = -x + x + 2$

$5 + 2 = \frac{1}{2}x - 2 + 2$ $\quad \frac{3}{2}x + 5 = 2$

$7 = \frac{1}{2}x$ $\quad \frac{3}{2}x + 5 - 5 = 2 - 5$

$\frac{2}{1} \cdot 7 = \frac{2}{1} \cdot \frac{1}{2}x$ $\quad \frac{2}{3} \cdot \frac{3}{2}x = -3 \cdot \frac{2}{3}$

$14 = x$ OR $x = -2$

Find the solutions of the following equations.

3. (21) $4|-s| = 24$

$|-s| = 6$

$-s = 6$ or $-s = -6$

$\{6, -6\}$

4. (41) $-3|x + 4| = -21$

$\frac{-3|x+4|}{-3} = \frac{-21}{-3}$

$|x+4| = 7$ or $|x+4| = -7$

$x = 7 - 4$ $\quad$ $x = -11$

$x = 3$

$\{3, -11\}$

5. $8 - 3|3x - 1| = -13$

$-3|3x-1| = -13 - 8$

$\frac{-3|3x-1|}{-3} = \frac{-21}{-3}$

$|3x-1| = 7$ or $|3x-1| = -7$

$\frac{3x}{3} = \frac{8}{3}$ $\quad$ $\frac{3x}{3} = \frac{-6}{3}$

$x = \frac{8}{3}$ $\quad$ $x = -2$

6. (69) $|1/2\ x + 5| = |x - 2|$

$\frac{1}{2}x + 5 = x - 2$ or $\frac{1}{2}x + 5 = -(x-2)$

$\frac{1}{2}x = x - 2 - 5$

$\frac{1}{2}x = x - 7$

$2(\frac{1}{2}x) = 2(x) - 2(7)$

$x = 2x - 14$

$-x = -14$

$x = 14$

$\frac{1}{2}x + 5 = -(x-2)$

$\frac{1}{2}x + 5 = -x + 2$

$2(\frac{1}{2}x) + 2(5) = 2(-x) + 2(2)$

$x + 10 = -2x + 4$

$\frac{3x}{3} = \frac{-6}{3}$

$x = -2$

$\{14, -2\}$

2·6 Absolute Value Inequalities

1. If an absolute value inequality is of the form $|ax + b|$ __<__ c, where c is a positive number, then only one region of the number line will be shaded.

2. Consider $|3x + 1| < 5$

$\underline{-5} < 3x + 1 < \underline{5}$

$\underline{-5-1} < 3x + 1 - 1 < \underline{5-1}$

$\underline{-6} < 3x < \underline{4}$

$\underline{\frac{1}{3}\cdot -6} < 1/3 \cdot 3x < \underline{\frac{1}{3}\cdot 4}$

$\underline{-2} < x < \underline{\frac{4}{3}}$

-3 -2 -1 0 1 4/3 2 3

3. If an absolute value inequality is of the form $|ax + b|$ __>__ c, where c is a positive number, then two distinct regions of the number line will be shaded.

4. Consider $|3x + 1| > 5$

$3x + 1 < -5$ or $3x + 1 > 5$

$3x + 1 - 1 < \underline{-5-1}$ or $3x + 1 - 1 > \underline{5-1}$

$3x < \underline{-6}$ or $3x > \underline{4}$

$1/3 \cdot 3x < \underline{\frac{1}{3}\cdot -6}$ or $1/3 \cdot 3x > \underline{\frac{1}{3}\cdot 4}$

$x < \underline{-2}$ or $x > \underline{\frac{4}{3}}$

-3 -2 -1 0 1 4/3 2

2·6 Absolute Value Inequalities

1. If an absolute value inequality is of the form $|ax + b|$ __<__ c, where c is a positive number, then only one region of the number line will be shaded.

2. Consider $|3x + 1| < 5$

   $\underline{-5} < 3x + 1 < \underline{5}$

   $\underline{-1-5} < 3x + 1 - 1 < \underline{5-1}$

   $\underline{-6} < 3x < \underline{4}$

   $\underline{-\frac{6}{3}} < 1/3 \cdot 3x < \underline{\frac{4}{3}}$

   $\underline{-2} < x < \underline{\frac{4}{3}}$

3. If an absolute value inequality is of the form $|ax + b|$ __>__ c, where c is a positive number, then two distinct regions of the number line will be shaded.

4. Consider $|3x + 1| > 5$

   $3x + 1 < -5$ or $3x + 1 > 5$

   $3x + 1 - 1 < \underline{-5-1}$ or $3x + 1 - 1 > \underline{5-1}$

   $3x < \underline{-6}$ or $3x > \underline{4}$

   $1/3 \cdot 3x < \underline{-\frac{6}{3}}$ or $1/3 \cdot 3x > \underline{\frac{4}{3}}$

   $x < \underline{-2}$ or $x > \underline{\frac{4}{3}}$

Find the solution of each of the following inequalities, and graph each solution on the number line.

5. $3 + |1/2\,x + 5| \le 3$

$3 - 3 + |\frac{1}{2}x + 5| \le 3 - 3$

$|\frac{1}{2}x + 5| \le 0$

$0 \le \frac{1}{2}x + 5 \le 0$

$-5 \le \frac{1}{2}x \le -5$

$-10 \le x \le -10$

$\boxed{x = -10}$

6. (23) $|x - 3| < 1/2$

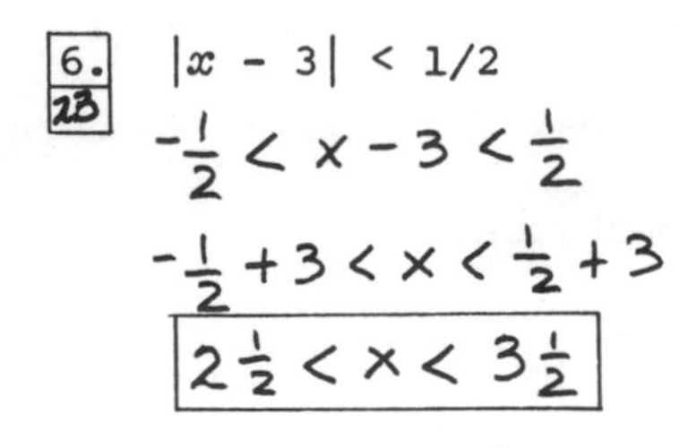

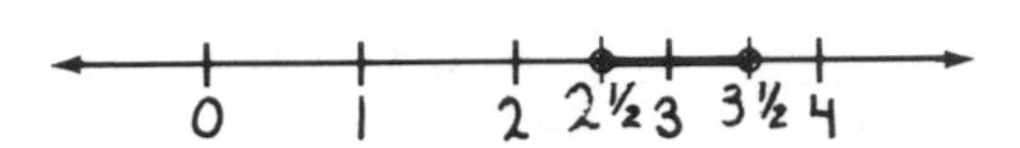

7. $|-2m + 1| > 9$

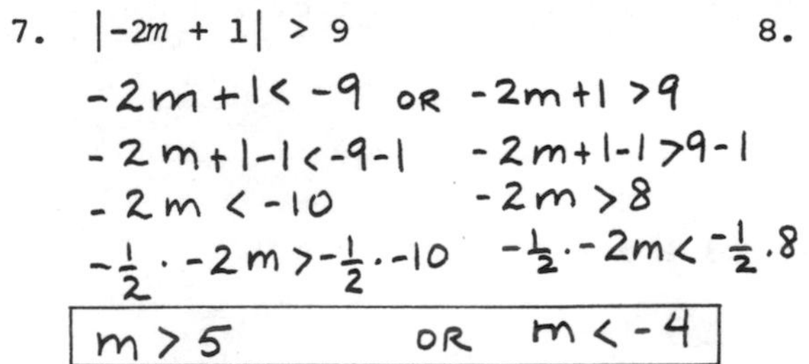

8. $|m| > 5$

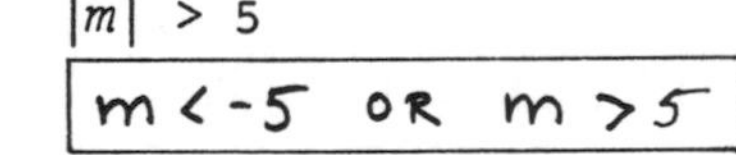

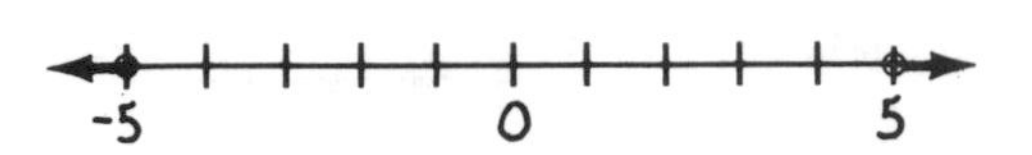

Find the solution of each of the following inequalities, and graph each solution on the number line.

5. $3 + \left|1/2\,x + 5\right| \leq 3$

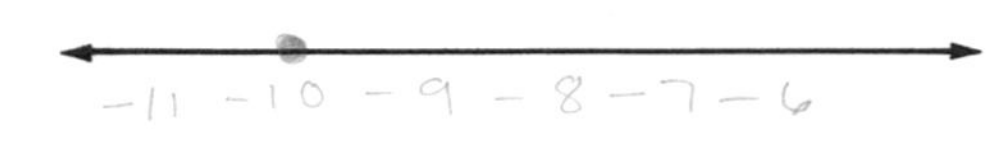

6. $|x - 3| < 1/2$

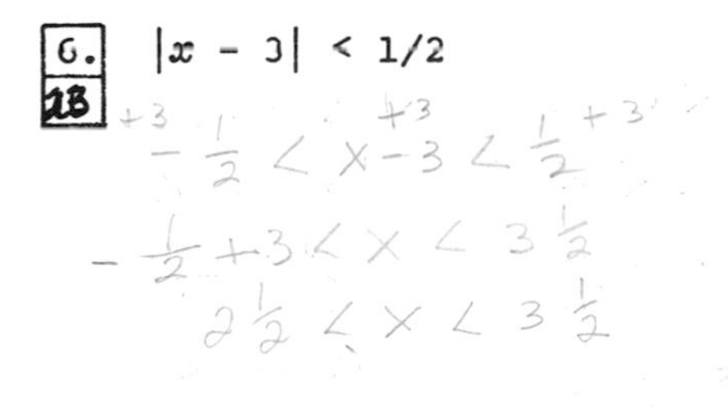

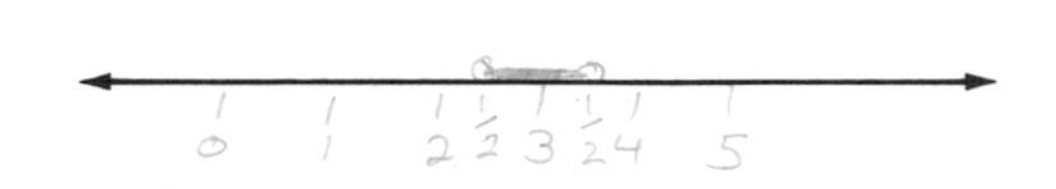

7. $|-2m + 1| > 9$

8. $|m| > 5$

9. $|7 - 3x| + 5 \geq 2$

$|7-3x|+5-5 \geq 2-5$

$|7-3x| \geq -3$

$(-1)|7-3x| \leq (-1)\cdot -3$

$-|7-3x| \leq 3$

$-(7-3x) \leq 3$ OR $-(7-3x) \geq -3$

$-7+3x \leq 3$ $\quad$ $-7+3x \geq -3$

$3x \leq 10$ $\quad$ $3x \geq 4$

$x \leq \frac{10}{3}$ $\quad$ $x \geq \frac{4}{3}$

$\boxed{\frac{4}{3} \leq x \leq \frac{10}{3}}$

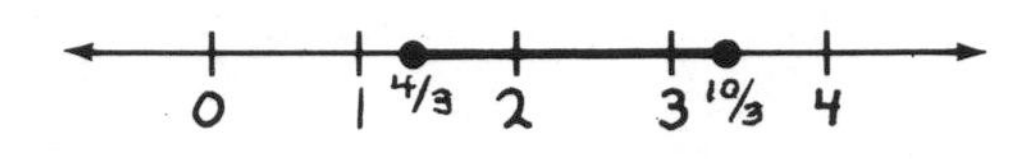

10. $2.5|2x - 1| + 1.8 < 9.3$

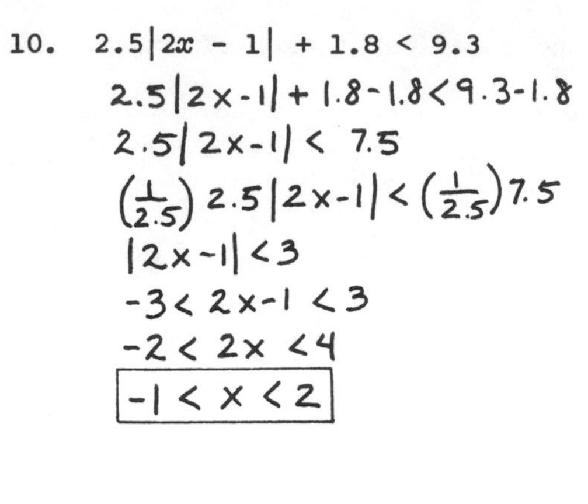

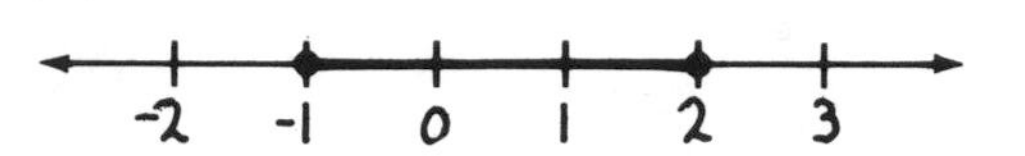

9. $|7 - 3x| + 5 \geq 2$

$|7-3x| \geq 2-5$

$|7-3x| \geq -3$

$(-1)|7-3x| \geq (-1)-3$

$-|7-3x| \leq 3$

$-(7-3x) \leq 3$

$-7+3x \leq 3$

$\frac{3x}{3} \leq \frac{10}{3}$

$x \leq \frac{10}{3}$

$-(7-3x) \geq -3$

$-7+3x \geq -3$

$\frac{3x}{3} \geq \frac{4}{3}$

$x \geq \frac{4}{3}$

10. $2.5|2x - 1| + 1.8 < 9.3$

-1.8

$2.5|2x-1| < 7.5$

$|2x-1| < 3$

$2x-1 < 3$

$\frac{2x}{2} < \frac{4}{2}$

$x < 2$

$2x-1 > -3$

$\frac{2x}{2} > \frac{-2}{2}$

$x > -1$

Chapter 2 Self-Test

Solve each equation or inequality. Graph the solutions to the inequalities.

1. $-5(x + 1) - (2x - 3) = -10$

   $-5x - 5 - 2x + 3 = -10$

   $-7x - 2 = -10$

   $\frac{-7x}{-7} = \frac{-8}{-7}$

   $x = \frac{8}{7}$

2. $2/3\ (x - 5) = x/6 + 1/2$

3. $.08 - 1.2x \leq .06x - 2.5$

4. $-5 < 3 - x \leq 8$

5. $2x + 3 > -5$ or $x - 1 \leq 8$

6. $-3x + 1 > -2$ and $x + 3 > 4$

7. $-3|x - 1| = -36$

8. $|1/4\ x + 3| = |x + 3|$

9. $|2x - 3| = -(x - 5)$

10. $|6 - 4x| + 8 \leq x + 14$

11. Solve $A = 1/2\ h(b_1 + b_2)$ for b_2.

12. Solve $V = 1/3\ \pi\ r^2 h$ for π.

13. Brenda has \$4.35 in dimes and nickels. She has a total of 62 coins. How many of each coin does she have?

14. John leaves Chicago at noon traveling west at a rate of 60 mph. Linda leaves from the same point at 3 p.m. traveling west at 75 mph. How many hours will it take Linda to overtake John?

CHAPTER

3 Polynomials and Factorable Quadratic Equations

3 · 1 Adding and Subtracting Polynomials

1. A polynomial is an algebraic expression containing a finite sum of terms.

2. Of the following: $\{\sqrt{3}\,x, \sqrt{3x}, 3x, 3/x, x + 3\}$, which are examples of polynomials?

   $\sqrt{3}\,x$, $3x$, $x+3$

3. A polynomial containing 2 terms is called a binomial; one with 1 term is called a monomial; one with 3 terms is called a trinomial. One with more than 3 terms is called no special name; just polynomial.

4. $1/2\, x^2 - 3x^4 + 15 - 3x$ is a polynomial in x.

   (a) Degree: 4 (b) Leading coefficient: -3

   (c) Standard form: $-3x^4 + \frac{1}{2}x^2 - 3x + 15$

CHAPTER

3 Polynomials and Factorable Quadratic Equations

3 · 1 Adding and Subtracting Polynomials

1. A ________________ is an algebraic expression containing a finite sum of terms.

2. Of the following: $\{\sqrt{3}\,x, \sqrt{3x}, 3x, 3/x, x + 3\}$, which are examples of polynomials?

   ____

3. A polynomial containing 2 terms is called a ______________; one with 1 term is called a ______________; one with 3 terms is called a ________________. One with more than 3 terms is called ____.

4. $1/2\, x^2 - 3x^4 + 15 - 3x$ is a polynomial in x.

   (a) Degree: _______ (b) Leading coefficient: ________

   (c) Standard form: ____________________________

5. $z^2 - 5xz^3 + 1/3\ x^2z$ is a polynomial in z. (Note that this tells us which term comes first.)

(a) Degree: 4 (b) Leading coefficient: -5

(c) Standard form: $-5xz^3 + z^2 + \frac{1}{3}x^2z$

Perform the indicated operations and express your answers in simplest form.

6. $5(3x - 2y)$

$5 \cdot 3x - 5 \cdot 2y =$

$\boxed{15x - 10y}$

7. $2(x + 3) + {-5}(3x - 1)$

$2x + 2 \cdot 3 + {-5} \cdot 3x + {-5} \cdot {-1} =$

$2x + 6 - 15x + 5 =$

$\boxed{-13x + 11}$

8. $.07[(2x^2 - 3x + 5) - (6x^2 + 5x - 2)]$

$.07[2x^2 - 3x + 5 - 6x^2 - 5x + 2] =$

$.07[-4x^2 - 8x + 7] = \boxed{-.28x^2 - .56x + .49}$

9. 77 $(5/6)[2a + (3/10)b] - (1/9)[5a - (3/2)b]$

$\left(\frac{5}{6}\right)(2a) + \left(\frac{5}{6}\right)\left(\frac{3}{10}\right)b - \left(\frac{1}{9}\right)(5a) + \left(\frac{1}{9}\right)\left(\frac{3}{2}\right)b =$

$\frac{10}{6}a + \frac{15}{60}b - \frac{5}{9}a + \frac{3}{18}b =$

$\left(\frac{30}{18}a - \frac{10}{18}a\right) + \left(\frac{3}{12}b + \frac{2}{12}b\right) =$

$\frac{20}{18}a + \frac{5}{12}b = \boxed{\frac{10}{9}a + \frac{5}{12}b}$

10. 57 $(x^2 - 5x + 6) - (3x^2 + x - 8) + (4x^2 - 13)$

$x^2 - 5x + 6 - 3x^2 - x + 8 + 4x^2 - 13 =$

$x^2 - 3x^2 + 4x^2 - 5x - x + 6 + 8 - 13 =$

$\boxed{2x^2 - 6x + 1}$

5. $z^2 - 5xz^3 + 1/3\ x^2z$ is a polynomial in z.

(a) Degree: ______ (b) Leading coefficient: ______

(c) Standard form: ____________________

Perform the indicated operations and express your answers in simplest form.

6. $5(3x - 2y)$

7. $2(x + 3) + -5(3x - 1)$

8. $.07[(2x^2 - 3x + 5) - (6x^2 + 5x - 2)]$

9. 77 $(5/6)[2a + (3/10)b] - (1/9)[5a - (3/2)b]$

10. 57 $(x^2 - 5x + 6) - (3x^2 + x - 8) + (4x^2 - 13)$

11. $(-1/2)(3x^3 - 4x + 6) + 8[(3/4)x^3 - (1/2)x^2 + 4x]$

$$\left[\left(-\frac{1}{2}\right)(3x^3) - \left(\frac{-1}{2}\right)(4x) + \left(\frac{-1}{2}\right)(6)\right] + \left[(8)\left(\frac{3}{4}\right)x^3 - (8)\left(\frac{1}{2}\right)x^2 + (8)(4x)\right] =$$

$$-\frac{3}{2}x^3 + 2x - 3 + 6x^3 - 4x^2 + 32x =$$

$$\boxed{\frac{9}{2}x^3 - 4x^2 + 34x - 3}$$

12. John has x nickels, y dimes, and z quarters. Find a polynomial that represents the value of his money.

The value of x nickels is $.05x$, since a nickel is worth \$.05.
The value of y dimes is $.10y$, since a dime is worth \$.10.
The value of z quarters is $.25z$, since a quarter is worth \$.25.
Therefore, the value of John's money is $\boxed{.05x + .10y + .25z}$

3 · 2 Multiplication of Polynomials

Write the correct choice in the blank provided.

Answer		Expression		Choice
c	1.	$(x^2)^3$	(a)	$-8x^6$
j	2.	$(x^2)(x^3)$	(b)	$6x^4 - 4x^2$
h	3.	$(xy)^2$	(c)	x^6
g	4.	$(x/2)^3$	(d)	$8x^9y^6 + 3x$
e	5.	$(x/2y)^2$	(e)	$x^2/(4y^2)$
i	6.	$(5x^2y^4)^2$	(f)	$6x^3 + 15x^5$
f	7.	$3x^2(2x + 5x^3)$	(g)	$x^3/8$
b	8.	$2x(3x^3 - 2x)$	(h)	x^2y^2
d	9.	$(2x^3y^2)^3 + 3x$	(i)	$25x^4y^8$
a	10.	$(-2x^2)^3$	(j)	x^5

11. $(-1/2)(3x^3 - 4x + 6) + 8[(3/4)x^3 - (1/2)x^2 + 4x]$

12. John has x nickels, y dimes, and z quarters. Find a polynomial that represents the value of his money.

3·2 Multiplication of Polynomials

Write the correct choice in the blank provided.

_____	1. $(x^2)^3$	(a) $-8x^6$
_____	2. $(x^2)(x^3)$	(b) $6x^4 - 4x^2$
_____	3. $(xy)^2$	(c) x^6
_____	4. $(x/2)^3$	(d) $8x^9y^6 + 3x$
_____	5. $(x/2y)^2$	(e) $x^2/(4y^2)$
_____	6. $(5x^2y^4)^2$	(f) $6x^3 + 15x^5$
_____	7. $3x^2(2x + 5x^3)$	(g) $x^3/8$
_____	8. $2x(3x^3 - 2x)$	(h) x^2y^2
_____	9. $(2x^3y^2)^3 + 3x$	(i) $25x^4y^8$
_____	10. $(-2x^2)^3$	(j) x^5

Perform the indicated multiplications. (Use the FOIL method if possible.)

11. $(3x + 2)(x - 5)$

$3x \cdot x - 3x \cdot 5 + 2x + 2 \cdot 5 =$

$3x^2 - 15x + 2x - 10 =$

$\boxed{3x^2 - 13x - 10}$

12. $(6m - 3k)(6m + 3k)$

$36m^2 + 18km - 18km - 9k^2 =$

$\boxed{36m^2 - 9k^2}$

13. $(x + 3)^2$

$(x+3)(x+3) =$

$x^2 + 3x + 3x + 9 =$

$\boxed{x^2 + 6x + 9}$

14. / 47 $(x/2 - 2y)(x/2 + 2y)$

$\left(\frac{x}{2}\right)\left(\frac{x}{2}\right) + \left(\frac{x}{2}\right)(2y) - (2y)\left(\frac{x}{2}\right) - (2y)(2y) =$

$\boxed{\frac{x^2}{4} - 4y^2}$

15. $[(2/3)a - 5b]^2$

$\left(\frac{2}{3}a - 5b\right)\left(\frac{2}{3}a - 5b\right) =$

$\frac{4}{9}a^2 - \frac{10}{3}ab - \frac{10}{3}ab + 25b^2 =$

$\boxed{\frac{4}{9}a^2 - \frac{20}{3}ab + 25b^2}$

16. $(p^4 - 1)^2$

$(p^4 - 1)(p^4 - 1) =$

$p^8 - p^4 - p^4 + 1 =$

$\boxed{p^8 - 2p^4 + 1}$

17. $(s^2 + 3)(s^3 - 5s^2 + 3)$

$s^2(s^3 - 5s^2 + 3) + 3(s^3 - 5s^2 + 3) =$

$s^5 - 5s^4 + 3s^2 + 3s^3 - 15s^2 + 9 =$

$\boxed{s^5 - 5s^4 + 3s^3 - 12s^2 + 9}$

18. / 73 $[(2y + 5)(y - 3)(2y + 1)]$

$(2y^2 - 6y + 5y - 15)(2y + 1) =$

$(2y^2 - y - 15)(2y + 1) =$

$4y^3 + 2y^2 - 2y^2 - y - 30y - 15 =$

$\boxed{4y^3 - 31y - 15}$

Perform the indicated multiplications. (Use the FOIL method if possible.)

11. $(3x + 2)(x - 5)$

12. $(6m - 3k)(6m + 3k)$

13. $(x + 3)^2$

14. 47 $(x/2 - 2y)(x/2 + 2y)$

15. $[(2/3)a - 5b]^2$

16. $(p^4 - 1)^2$

17. $(s^2 + 3)(s^3 - 5s^2 + 3)$

18. 73 $(2y + 5)(y - 3)(2y + 1)$

Identify each of the following as

(a) a perfect square trinomial or

(b) the difference of two squares.

b 19. $a^2 - b^2$ $(a+b)(a-b)$

b 20. $(3x - 1)(3x + 1)$ $9x^2-1$

a 21. $k^2 - 4k + 4$ $(K-2)(K-2) = (K-2)^2$

b 22. $16x^2 - 1$ $(4x+1)(4x-1)$

a 23. $(3x - 2)^2$ $9x^2-12x+4$

a 24. $9y^2 - 30y + 25$ $(3y-5)(3y-5) = (3y-5)^2$

3.3 Factoring Polynomials—Greatest Common Factor

1. Writing a polynomial as a product is called factoring the polynomial.

2. Find the Greatest Common Factor (GCF) of $12x^3$ and $18x^4$.

To find their greatest common factor:

$12x^3 = 2 \cdot 2 \cdot 3 \cdot x \; x \; x$

$18x^4 = 2 \cdot 3 \; 3 \; x \; x \; x \cdot x$

GCF = $2 \cdot 3 \cdot x \cdot x \cdot x$

$= 6 \quad x^3$

$= 6x^3$

By writing the prime factorizations in this manner, it is easy to see that the common factors are 2, 3, x, x, x. Therefore, the GCF is $6x^3$.

The factorization of $12x^3 + 18x^4 = 6x^3(2) + 6x^3(3x)$ This is the GCF!

$= 6x^3(2 + 3x)$

Identify each of the following as

(a) a perfect square trinomial or

(b) the difference of two squares.

_____ 19. $a^2 - b^2$

_____ 20. $(3x - 1)(3x + 1)$

_____ 21. $k^2 - 4k + 4$

_____ 22. $16x^2 - 1$

_____ 23. $(3x - 2)^2$

_____ 24. $9y^2 - 30y + 25$

3.3 Factoring Polynomials—Greatest Common Factor

1. Writing a polynomial as a product is called <u>factoring</u> the polynomial.

2. Find the Greatest Common Factor (GCF) of $12x^3$ and $18x^4$.

   To find their greatest common factor:

$$12x^3 = 2 \cdot 2 \cdot 3 \cdot x \cdot x \cdot x$$

$$18x^4 = 2 \cdot 3 \cdot 3 \cdot x \cdot x \cdot x \cdot x$$

$$\text{GCF} = 2 \cdot 3 \cdot x \cdot x \cdot x$$

$6x^3$

By writing the prime factorizations in this manner, it is easy to see that the common factors are 2, 3, x, x, x. Therefore, the GCF is $6x^3$.

The factorization of $12x^3 + 18x^4 = 6x^3(2) + 6x^3(3x)$

$$= 6x^3(2 + 3x)$$

(a) Find the GCF of $18x^3y$ and $54x^2y^2$.

$18x^3y = 2 \cdot 3 \cdot 3 \cdot \quad \cdot x \cdot x \cdot x \cdot y$

$54x^2y^2 = 2 \cdot 3 \cdot 3 \cdot 3 \cdot x \cdot x \cdot \quad y \cdot y$

GCF = $2 \cdot 3 \cdot 3 \cdot x \cdot x \cdot y$ = $18x^2y$ = GCF

(b) Factor: $18x^3y - 54x^2y^2$

$18x^3y - 54x^2y^2 = 18x^2y(x - 3y)$

(c) Find the GCF of $3x^4y^2$, $9x^3y^2$, and $3x^2y^2$.

$3x^4y^2 = 3 \cdot \quad \cdot x \cdot x \cdot x \cdot x \cdot y \cdot y$

$9x^3y^2 = 3 \cdot 3 \cdot x \cdot x \cdot x \cdot y \cdot y$

$3x^2y^2 = 3 \cdot \quad \cdot x \cdot x \cdot y \cdot y$

GCF = $3x^2y^2$

(d) Factor: $3x^4y^2 + 9x^3y^2 + 3x^2y^2 =$

$3x^2y^2(x^2 + 3x + 1)$

3. Factor the following polynomials.

(a) 33 $4n + 15mn - 3n^2 - 20m$

(b) $6a + 4a^2 - 6ab + 3b$

(c) $3x^3 + 5y + 3x^2y - xy$

(d) 23 $3x^2(x + y)^2 + 12x(x + y)$

(a) Find the GCF of $18x^3y$ and $54x^2y^2$.

$18x^3y = 2 \cdot 3 \cdot 3 \cdot x \cdot x \cdot x \cdot y$

$54x^2y^2 = 2 \cdot 3 \cdot 3 \cdot 3 \cdot x \cdot x \cdot y \cdot y$

GCF = $18x^2y$

(b) Factor: $18x^3y - 54x^2y^2$

$18x^2y(x - 3y)$

(c) Find the GCF of $3x^4y^2$, $9x^3y^2$, and $3x^2y^2$.

$3x^4y^2 = 3 \cdot x \cdot x \cdot x \cdot x \cdot y \cdot y$

$9x^3y^2 = 3 \cdot 3 \cdot x \cdot x \cdot x \cdot y \cdot y$

$3x^2y^2 = 3 \cdot x \cdot x \cdot y \cdot y$

GCF = $3x^2y^2$

(d) Factor: $3x^4y^2 + 9x^3y^2 + 3x^2y^2$

$3x^2y^2(x^2 + 3x + 1)$

3. Factor the following polynomials.

(a)
33

$4n + 15mn - 3n^2 - 20m$

(b) $6a + 4a^2 - 6ab + 3b$

(c) $3x^3 + 5y + 3x^2y - xy$

(d)
23

$3x^2(x + y)^2 + 12x(x + y)$

3.4 Factoring Binomials

1. Factor as the difference of two squares, if possible. If not possible, write "Prime."

(a) $16a^2 - 1 = (4a+1)(4a-1)$

(b) $3x^2 - 48 = 3(x^2-16) = 3(x+4)(x-4)$

(c) $x^2 - 7$ Prime

(d) $.04x^2 - 1.21 = (.2x+1.1)(.2x-1.1)$

(e) $49x^2 - 21$ Prime

(f) $z^4 - 81 = (z^2+9)(z^2-9)$
$= (z^2+9)(z+3)(z-3)$

(g) $4x^2 + 25y^2$ Prime

(h) $(x + 2)^2 - 100 = [(x+2)+10][(x+2)-10]$
$= (x+12)(x-8)$

2. (a) $x^3 + y^3 = (x + y)(x^2 - xy + y^2)$

(b) $x^3 - y^3 = (x - y)(x^2 + xy + y^2)$

Factor each polynomial completely.

3. $5x^3 - 135y^3 =$
$5(x^3 - 27y^3) =$
$5(x-3y)(x^2+3xy+9y^2)$

4. $x^3 + 8 =$
$(x+2)(x^2-2x+4)$

5. $(3a - 1)^3 - 27 =$
$[(3a-1)-3][(3a-1)^2+3(3a-1)+9] =$
$(3a-4)[9a^2-6a+1+9a-3+9] =$
$(3a-4)(9a^2+3a+7)$

6. (36) $(3y + 2)^3 + (y - 1)^3$
$= [(3y+2)+(y-1)][(3y+2)^2 - (3y+2)(y-1) + (y-1)^2]$
$= (4y+1)(9y^2+12y+4-3y^2+y+2+y^2-2y+1)$
$= (4y+1)(7y^2+11y+7)$

3.4 Factoring Binomials

1. Factor as the difference of two squares, if possible. If not possible, write "Prime."

(a) $16a^2 - 1$ $(4a-1)(4a+1)$

(b) $3x^2 - 48$ $3(x^2-16)$ $3(x-4)(x+4)$

(c) $x^2 - 7$ prime

(d) $.04x^2 - 1.21$ $(.2x-1.1)(.2x+1.1)$

(e) $49x^2 - 21$ prime

(f) $z^4 - 81$ $(z^2-9)(z^2+9)$ $(z-3)(z+3)(z^2+9)$

(g) $4x^2 + 25y^2$ prime

(h) $(x + 2)^2 - 100$ $[(x+2)-10][(x+2+10)]$ $(x-8)(x+12)$

2. (a) $x^3 + y^3 = (x \underline{+} y)(x^2 \underline{-} xy \underline{+} y^2)$

(b) $x^3 - y^3 = (x \underline{-} y)(x^2 \underline{+} xy \underline{+} y^2)$

Factor each polynomial completely.

3. $5x^3 - 135y^3$

$5(x^3 - 27y^3)$

$5(x-3y)(x^2+3xy+9y^2)$

4. $x^3 + 8$

$x+2(x^2-2x+4)$

5. $(3a - 1)^3 - 27$

$(3a-1) - 3$

$(3a-1-3)[(3a-1)^2 \; 3(3a-1) \; 9]$

$3a-4[(3a-1)(3a-1)+9a-3+9]$

$(3a-4)(9a^2-6a+1)+9a-3+9)$

$3a-4(9a^2+9a-6a-3+1+9)$

$(3a-4)(9a^2+3a+7)$

6. [36] $(3y + 2)^3 + (y - 1)^3$

3.5 Factoring Trinomials

1\. [21] $2x^3 + 16x^2 + 32x =$

$2x(x^2+8x+16) =$

$2x(x+4)(x+4) =$

$\boxed{2x(x+4)^2}$

2\. $x^2 - 5x + 6 =$

$\boxed{(x-3)(x-2)}$

3\. $3x^2 - 14xy - 5y^2$

$\boxed{(3x+y)(x-5y)}$

4\. $12x^2 + 20x - 8 =$

$4(3x^2+5x-2) =$

$\boxed{4(3x-1)(x+2)}$

5\. [64] $8x^2 + 6xy - 5y^2 =$

$\boxed{(4x+5y)(2x-y)}$

$8x^2 - 4xy + 10xy - 5y^2$

6\. $x^2 + 6x - 5$

Not factorable;

$\boxed{\text{prime}}$

7\. $6x^3 - 9x^2y + 2xy - 3y^2 =$

$(6x^3 - 9x^2y) + (2xy - 3y^2) =$

$3x^2(2x-3y) + y(2x-3y) =$

$\boxed{(3x^2+y)(2x-3y)}$

8\. [58] $16x^4 - 44x^3 + 24x^2 =$

$4x^2(4x^2 - 11x + 6) =$

$\boxed{4x^2(4x-3)(x-2)}$

9\. $m^2 + 5mn + 6n^2$

$\boxed{(m+3n)(m+2n)}$

10\. $2x^4 - 2x^2 - 24 =$

$2(x^4 - x^2 - 12) =$

$2(x^2-4)(x^2+3) =$

$\boxed{2(x+2)(x-2)(x^2+3)}$

11\. $x^2 - 6x + 9 - 4y^2 =$

$(x^2-6x+9) - 4y^2 =$

$(x-3)^2 - 4y^2 =$

$\boxed{[(x-3)+2y][(x-3)-2y]}$

12\. [79] $6(x + 2y)^2 - 19(x + 2y) + 8$

$= [3(x+2y)-8][2(x+2y)-1]$

$= \boxed{(3x+6y-8)(2x+4y-1)}$

3.5 Factoring Trinomials

1. $2x^3 + 16x^2 + 32x$

   21

   $2x(x^2 + 8x + 16)$

   $2x(x + 4)(x + 4)$

2. $x^2 - 5x + 6$

   $(x - 2)(x - 3)$

3. $3x^2 - 14xy - 5y^2$

   $(3x + y)(x - 5y)$

4. $12x^2 + 20x - 8$

   $4(3x^2 + 5x - 2)$

   $(3x - 1)(x + 2)$

5. $8x^2 + 6xy - 5y^2$

   64

6. $x^2 + 6x - 5$

7. $6x^3 - 9x^2y + 2xy - 3y^2$

   $(6x^3 - 9x^2y) + (2xy - 3y^2)$

   $3x^2(2x - 3y) + y(2x - 3y)$

   $(3x^2 + y)(2x - 3y)$

8. $16x^4 - 44x^3 + 24x^2$

   58

   $4x^2(4x^2 - 11x + 6)$

   $4x^2(4x - 3)(x - 2)$

9. $m^2 + 5mn + 6n^2$

   $(m + 2n)(m + 3n)$

10. $2x^4 - 2x^2 - 24$

    $2(x^2 + 3)(x^2 - 4)$

11. $x^2 - 6x + 9 - 4y^2$

12. $6(x + 2y)^2 - 19(x + 2y) + 8$

    79

3.6 Quadratic Equations

1. (a) The standard form of a quadratic equation in the variable x is:

   $ax^2 + bx + c = 0$.

   (b) Express the following in standard form: $3x - 5 = 2x^2$

   $3x - 5 = 2x^2$

   $-2x^2 + 3x - 5 = 0$ OR $2x^2 - 3x + 5 = 0$

Find solutions to the following quadratic equations.

2. $15x^2 + x = 6$

   $15x^2 + x - \underline{6} = \underline{0}$ (1) Express in standard form.

   $(3x + \underline{2})(5x - \underline{3}) = \underline{0}$ (2) Factor.

   $(3x + \underline{2}) = 0$ or $(5x - \underline{3}) = 0$ (3) Set each factor equal to 0 (since if $ab = 0$ either $a = 0$ or $b = 0$).

   $3x = \underline{-2}$ or $5x = \underline{3}$ (4) Solve for x.

   $1/3 \cdot 3x = \underline{-2 \cdot \frac{1}{3}}$ or $1/5 \cdot 5x = \underline{3 \cdot \frac{1}{5}}$

   $x = \underline{-\frac{2}{3}}$ or $x = \underline{\frac{3}{5}}$

   $15(\underline{-\frac{2}{3}})^2 + \underline{-\frac{2}{3}} = 6$ or $15(\underline{\frac{3}{5}})^2 + \underline{\frac{3}{5}} = 6$ (5) Check solutions.

   $\underline{15(\frac{4}{9}) + -\frac{2}{3}} = 6$ $\underline{15(\frac{9}{25}) + \frac{3}{5}} = 6$

   $\underline{\frac{60}{9} - \frac{6}{9}} = 6$ $\underline{\frac{27}{5} + \frac{3}{5}} = 6$

   $6 = 6$ $6 = 6$

3.6 Quadratic Equations

1. (a) The standard form of a quadratic equation in the variable x is:

$ax^2 + bx + c = 0$.

(b) Express the following in standard form: $3x - 5 = 2x^2$

$-2x^2 + 3x - 5 = 0$

or $2x^2 - 3x + 5 = 0$

Find solutions to the following quadratic equations.

2. $15x^2 + x = 6$

$15x^2 + x - \underline{6} = \underline{0}$ (1) Express in standard form.

$(3x + \underline{2})(5x - \underline{3}) = \underline{0}$ (2) Factor.

$(3x + \underline{2}) = 0$ or $(5x - \underline{3}) = 0$ (3) Set each factor equal to 0 (since if $ab = 0$ either $a = 0$ or $b = 0$).

$3x = \underline{-2}$ or $5x = \underline{3}$ (4) Solve for x.

$1/3 \cdot 3x = \underline{-\frac{2}{3}}$ or $1/5 \cdot 5x = \underline{\frac{3}{5}}$

$x = \_\_\_\_$ or $x = \_\_\_\_$

$15(\_\_\_)^2 + \_\_\_ = 6$ or $15(\_\_\_)^2 + \_\_\_ = 6$ (5) Check solutions.

$\_\_\_\_\_\_ = 6$ $\_\_\_\_\_\_ = 6$

$\_\_\_\_\_\_ = 6$ $\_\_\_\_\_\_ = 6$

$6 = 6$ $6 = 6$

3. $-100y^2 + 36 = 0$

$(-1)(-100y^2+36) = (-1)(0)$

$100y^2 - 36 = 0$

$(10y+6)(10y-6) = 0$

$10y+6 = 0$ OR $10y-6 = 0$

$10y = -6$ $\quad$ $10y = 6$

$y = \frac{-6}{10}$ $\quad$ $y = \frac{6}{10}$

$y = \frac{-3}{5}$ OR $y = \frac{3}{5}$

4. $7x^2 = 2x$

$7x^2 - 2x = 0$

$x(7x-2) = 0$

$x = 0$ OR $7x-2 = 0$

$7x = 2$

$x = 0$ OR $x = \frac{2}{7}$

5. | 39 |

$8x^2 + 19x + 6 = 0$

$(8x+3)(x+2) = 0$

$8x+3 = 0$ OR $x+2 = 0$

$8x = -3$

$x = \frac{-3}{8}$ OR $x = -2$

6. | 53 |

$x^2 + 5x - 4 = 10$

$(x^2+5x-4-10) = 0$

$x^2+5x-14 = 0$

$(x+7)(x-2) = 0$

$x+7 = 0$ OR $x-2 = 0$

$x = -7$ OR $x = 2$

7. $5x^2 - 11x - 4 = 1 - x^2 - 24x$

$5x^2 - 11x - 4 - 1 + x^2 + 24x = 0$

$6x^2 + 13x - 5 = 0$

$(2x+5)(3x-1) = 0$

$2x+5 = 0$ OR $3x-1 = 0$

$2x = -5$ $\quad$ $3x = 1$

$x = -\frac{5}{2}$ OR $x = \frac{1}{3}$

8. $x^2 - 6x + 12 = 2(x - 2)$

$x^2 - 6x + 12 = 2x - 4$

$x^2 - 6x + 12 - 2x + 4 = 0$

$x^2 - 8x + 16 = 0$

$(x-4)(x-4) = 0$

$x-4 = 0$ OR $x-4 = 0$

$x = 4$ OR $x = 4$

$10 \cdot 10$ $\quad 6 \cdot 6$

3. $-100y^2 + 36 = 0$

$-100y^2 + 0y + 36 = 0$

$100y^2 - 0y - 36 = 0$

$(10y - 6)(10y + 6) = 0$

$+\frac{6}{10}, -\frac{6}{10}$

$\{\frac{3}{5}, -\frac{3}{5}\}$

4. $7x^2 = 2x$

$7x^2 - 2x = 0$

$x(7x - 2) = 0$

$\{0, \frac{2}{7}\}$

5. 39 $\quad 8x^2 + 19x + 6 = 0$

$(8x + 3)(x + 2) = 0$

$\{-\frac{3}{8}, -2\}$

6. 53 $\quad x^2 + 5x - 4 = 10$

$x^2 + 5x - 4 - 10 = 0$

$x^2 + 5x - 14 = 0$

$(x - 2)(x + 7) = 0$

$\{2, -7\}$

7. $5x^2 - 11x - 4 = 1 - x^2 - 24x$ $\quad -1 \quad +x^2 \quad +24x$

$5x^2 + x^2 - 11x + 24x - 4 - 1$

$6x^2 + 13x - 5 = 0$

$(3x - 1)(2x + 5) = 0$

$\{\frac{1}{3}, -\frac{5}{2}\}$

8. $x^2 - 6x + 12 = 2(x - 2)$

$x^2 - 6x + 12 = 2x - 4$

$x^2 - 6x - 2x + 12 + 4 = 0$

$x^2 - 8x + 16 = 0$

$(x - 4)(x - 4) = 0$

$\{4\}$

9. $2x^2 + .05 = .7x$

$2x^2 - .7x + .05 = 0$

$(2x - .5)(x - .1) = 0$

$2x - .5 = 0$ OR $x - .1 = 0$

$2x = .5$ $\quad$ $x = .1$

$x = .25$ OR $x = .1$

10. $1/4\, x^2 + x - 15 = 0$

$(\frac{1}{2}x + 5)(\frac{1}{2}x - 3) = 0$

$\frac{1}{2}x + 5 = 0$ OR $\frac{1}{2}x - 3 = 0$

$\frac{1}{2}x = -5$ $\quad$ $\frac{1}{2}x = 3$

$x = -10$ OR $x = 6$

11. (73) $(2x + 5)(x - 4) = (x + 7)(x - 4)$

$2x^2 - 3x - 20 = x^2 + 3x - 28$

$2x^2 - 3x - 20 - x^2 - 3x + 28 = 0$

$x^2 - 6x + 8 = 0$

$(x - 2)(x - 4) = 0$

$x - 2 = 0$ OR $x - 4 = 0$

$x = 2$ OR $x = 4$

3.7 Applications

Use quadratic equations to find solutions of the following problems:

1. (5) The sum of the squares of two consecutive positive integers is 25.

Find the integers.

x = smaller integer

$x+1$ = larger integer

$(x)^2 + (x+1)^2 = 25$

$x^2 + 2x + 1 = 25$

$2x^2 + 2x - 24 = 0$

$2(x^2 + x - 12) = 0$

$2(x+4)(x-3) = 0$

$x + 4 = 0$ OR $x - 3 = 0$

$x = -4$ OR $x = 3$

These are solutions of the equation, not the stated problem.
* Since the problem specifies positive integers, we discard −4 as a solution. So:

$x = 3$ $\quad$ $x + 1 = 4$

9. $2x^2 + .05 = .7x$

$2x^2 - .7 + .05 = 0$

$(2x - .05)(x - .1)$

$\{.025, 1\}$

or $(2x - .5)(x - .1)$

$\{.25, .1\}$

10. $1/4\,x^2 + x - 15 = 0$

$(\frac{1}{2}x + 5)(\frac{1}{2}x - 3)$

$\frac{5}{\frac{1}{2}}$ $\frac{3}{\frac{1}{2}}$

$\{-10, 6\}$

or $4(\frac{1}{4}x^2) + 4(x) - 4(15) = 0$

$x^2 + 4x - 60 = 0$

$(x - 6)(x + 10) = 0$

$\{6, -10\}$

11. (73) $(2x + 5)(x - 4) = (x + 7)(x - 4)$

$2x^2 - 8x + 5x - 20 = x^2 - 4x + 7x - 28$

$2x^2 - 3x - 20 = x^2 + 3x - 28$

$-x^2 \quad -3x \quad +28$

$x^2 - 6x + 8 = 0$

$(x - 2)(x - 4) = 0$

$\{2, 4\}$

3.7 Applications

Use quadratic equations to find solutions of the following problems:

1. (5) The sum of the squares of two consecutive positive integers is 25. Find the integers.

$x^2 + (x+1)^2 = 25$

$(x+1)(x+1) + x^2 = 25$

$x^2 + 2x + 1 + x^2 = 25$

$x^2 + x^2 + 2x + 1 = 25$

$2x^2 + 2x + 1 - 25 = 0$

$2x^2 + 2x - 24 = 0$

$2(x^2 + x - 12) = 0$

$2(x - 3)(x + 4) = 0$

3 or -4

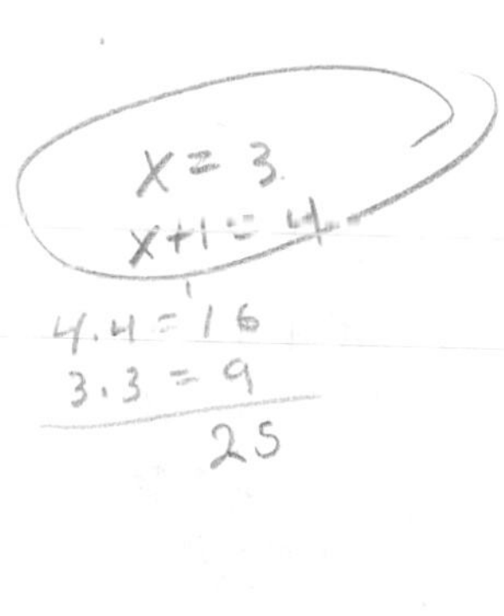

2. **25**

A painting and its frame cover 117 square inches. The frame is 3/2 inches wide. The length of the painting is 4 inches greater than the width. What are the dimensions of the painting?

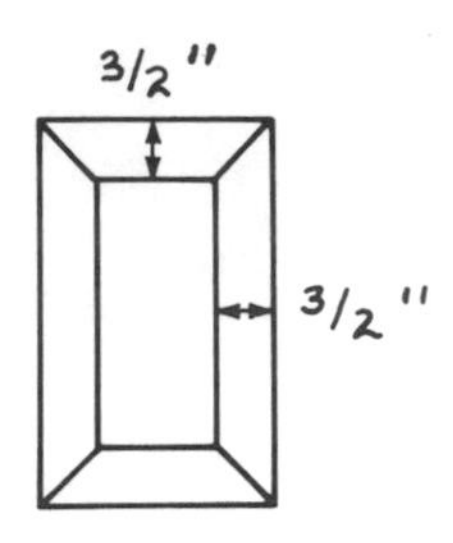

x = width of painting

$x + \frac{3}{2} + \frac{3}{2} = x + 3$ = width of painting and frame

$x + 4$ = length of painting

$x + 4 + \frac{3}{2} + \frac{3}{2}$ = length of painting and frame

$= x + 7$

$L \times W = \text{Area}$

$(x+7)(x+3) = 117$

$x^2 + 10x + 21 = 117$

$x^2 + 10x - 96 = 0$

$(x+16)(x-6) = 0$

$x + 16 = 0$ or $x - 6 = 0$

$x = -16$

* We discard this solution because a painting cannot have a negative width.

$x = 6$" WIDTH
$x + 4 = 10$" LENGTH

3. **13**

One leg of a right triangle is 2 inches longer than the other leg. The hypotenuse is 10 inches long. What are the lengths of the legs of the triangle? (Hint: Use the Pythagorean Theorem.)

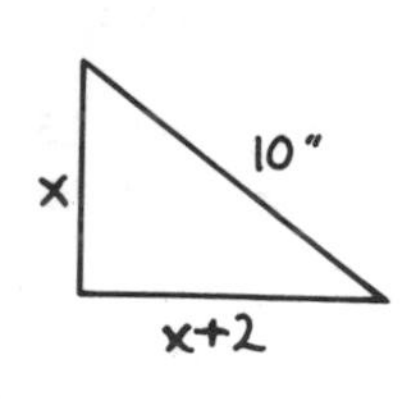

x = length of shorter leg

$x + 2$ = length of longer leg

$a^2 + b^2 = c^2$ (Pythagorean Theorem)

$x^2 + (x+2)^2 = 10^2$

$x^2 + x^2 + 4x + 4 = 100$

$2x^2 + 4x - 96 = 0$

$x^2 + 2x - 48 = 0$

$(x+8)(x-6) = 0$

$x + 8 = 0$ or $x - 6 = 0$

$x = -8$ $\quad$ $x = 6$

* Length cannot be negative so discard -8

$x = 6$ inches is length of shorter leg
$x + 2 = 8$ inches is length of longer leg

2. 25

A painting and its frame cover 117 square inches. The frame is 3/2 inches wide. The length of the painting is 4 inches greater than the width. What are the dimensions of the painting?

3. 13

One leg of a right triangle is 2 inches longer than the other leg. The hypotenuse is 10 inches long. What are the lengths of the legs of the triangle? (Hint: Use the Pythagorean Theorem.)

4. 17 A bullet is fired from ground level vertically upward with a speed of 384 ft/sec. The equation that gives us the bullet's height above ground level is $h = -16t^2 + 384t$. (1) How high is the bullet after 5 seconds? (2) How high is the bullet after 12 seconds? (3) When does it hit the ground?

(1) $h = -16(5)^2 + 384(5)$

$h = -400 + 1920$

$\boxed{h = 1520 \text{ feet}}$

(2) $h = -16(12)^2 + 384(12)$

$h = -2304 + 4608$

$\boxed{h = 2304 \text{ feet}}$

(3) When bullet hits the ground, $h = 0$.

$0 = -16t^2 + 384t$

$0 = -16t(t - 24)$

$-16t = 0$ OR $t - 24 = 0$

$t = 0$ sec. $\boxed{t = 24 \text{ sec.}}$

5. 23 Charley and Loretta have a rectangular swimming pool that is 8 feet wide and 14 feet long. They are going to build a tile border around the pool of uniform width. They have 75 square feet of tile. How wide is the border?

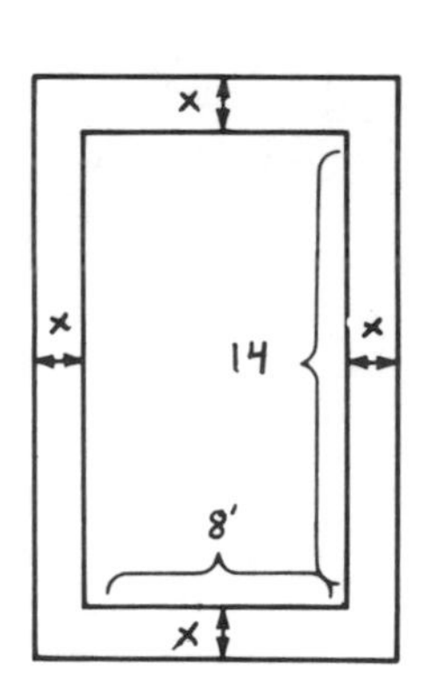

width of pool plus border = $2x + 8$

Length of pool plus border = $2x + 14$

Total area − Area of pool = area of border

$(2x+8)(2x+14) - (8)(14) = 75$

$4x^2 + 44x + 112 - 112 = 75$

$4x^2 + 44x - 75 = 0$

$(2x - 3)(2x + 25) = 0$

$2x - 3 = 0$ OR $2x + 25 = 0$

$x = \frac{3}{2}$ OR $x = \frac{-25}{2}$

Since the problem concerns dimension, we choose:

$x = \frac{3}{2}$ OR $\boxed{x = 1\frac{1}{2} \text{ feet is border width}}$

4.
17

A bullet is fired from ground level vertically upward with a speed of 384 ft/sec. The equation that gives us the bullet's height above ground level is $h = -16t^2 + 384t$. ① How high is the bullet after 5 seconds? ② How high is the bullet after 12 seconds? ③ When does it hit the ground?

5.
23

Charley and Loretta have a rectangular swimming pool that is 8 feet wide and 14 feet long. They are going to build a tile border around the pool of uniform width. They have 75 square feet of tile. How wide is the border?

6. | 9

Find two numbers whose difference is 7 and whose product is -12.

x = smaller number

$x+7$ = larger number

$$x(x+7) = -12$$
$$x^2+7x = -12$$
$$x^2+7x+12 = 0$$
$$(x+4)(x+3) = 0$$
$$x+4=0 \quad \text{OR} \quad x+3=0$$
$$x=-4 \quad \text{OR} \quad x=-3$$
$$x+7=3 \qquad x+7=4$$

-4 and 3 OR -3 and 4

6. 9 Find two numbers whose difference is 7 and whose product is -12.

Chapter 3 Self-Test

Perform the indicated operations. Express your answer in its simplest form.

1. $3(x + 5) - 6(2x - 1)$

2. $(2x - 1)(x + 5)$

3. $.03[(3x^2 - 4x) - (6x^2 - 2x + 5)]$

4. $(a^3 - 4)^2$

5. $(1/3\ x + 1/2\ y)(2/3\ x - 1/4\ y)$

6. $(p^2 - 1)(p^3 + 3p^2 - 5p + 2)$

Factor each expression completely, if possible.

7. $4x^3y^2 - 24x^5y^4 + 12y^2$

8. $(x - 8)^2 - 144$

9. $.04x^2 + 1.21$

10. $6x^2 - xy - 12y^2$

11. $27x^3 - 1$

12. $2ac + 3bc - 2ad - 3bd$

Solve by factoring.

13. $1/4\ x^2 + x - 15 = 0$

14. $.25x^2 + 2.5x = -6.25$

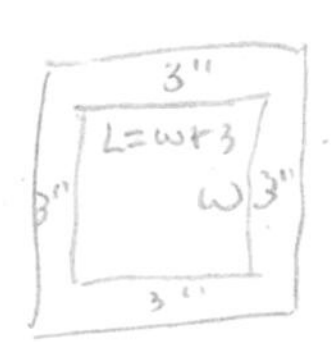

15. A painting and its frame cover 154 square inches. The frame is 3 inches wide. The length of the painting is 3 inches greater than the width. What are the dimensions of the painting?

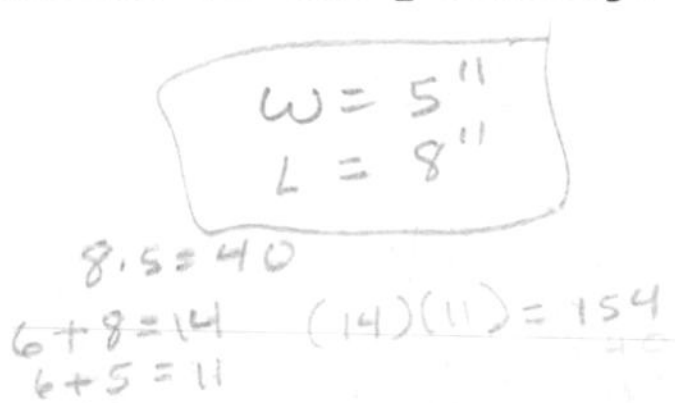

16. One leg of a right triangle is 14 inches longer than the other. The hypotenuse is 26 inches long. What are the lengths of the legs of the triangle?

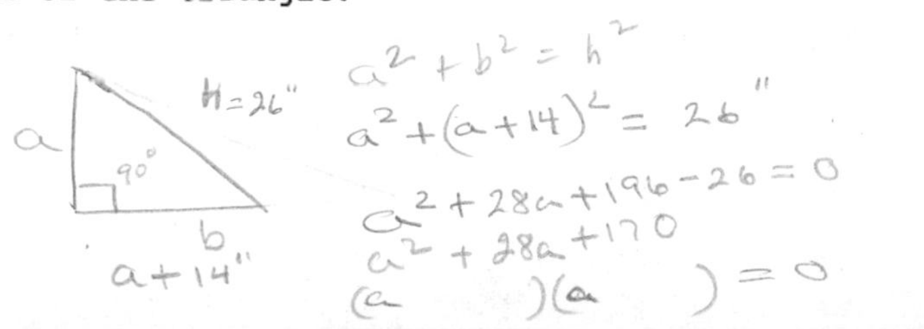

C H A P T E R

4 Rational Expressions

4 · 1 Integer Exponents

Write the correct choice in the blank provided.

c	1. $\frac{a^x}{a^y}$	(a)	$a^x b^x$
e	2. $\left(\frac{a}{b}\right)^x$	(b)	$\frac{1}{a^x b^y}$
l	3. a^{-y}	(c)	a^{x-y}
d	4. $a^x a^y$	(d)	a^{x+y}
g	5. $\frac{a^y}{a^x}$	(e)	$\frac{a^x}{b^x}$
i	6. $(a^x)^y$	(f)	-1
a	7. $(ab)^x$	(g)	a^{y-x}
h	8. a^0	(h)	1
f	9. $-a^0$	(i)	a^{xy}
k	10. a^{-1}	(j)	$\frac{b}{a}$

C H A P T E R

4 Rational Expressions

4·1 Integer Exponents

Write the correct choice in the blank provided.

Answer	Expression	Choice
c	1. $\frac{a^x}{a^y}$	(a) $a^x b^x$
e	2. $\left(\frac{a}{b}\right)^x$	(b) $\frac{1}{a^x b^y}$
l	3. a^{-y}	(c) a^{x-y}
d	4. $a^x a^y$	(d) a^{x+y}
g	5. $\frac{a^y}{a^x}$	(e) $\frac{a^x}{b^x}$
i	6. $(a^x)^y$	(f) -1
a	7. $(ab)^x$	(g) a^{y-x}
h	8. a^0	(h) 1
f	9. $-a^0$	(i) a^{xy}
k	10. a^{-1}	(j) $\frac{b}{a}$

j 11. $\left(\frac{a}{b}\right)^{-1}$ (k) $\frac{1}{a}$

b 12. $a^{-x}b^{-y}$ (l) $\frac{1}{a^y}$

Simplify each of the following. Remember to write your answers with positive exponents only.

13. $(4xy^3)(8x^3y^4)^2$

$4xy^3 \cdot 64x^6y^8 =$

$256x^7y^{11}$

14. $\left(\frac{2}{5}\right)^{-3} = \left(\frac{5}{2}\right)^3$

$= \frac{125}{8}$

15. 23 $4^{-1} + 2^{-2}$

$\frac{1}{4} + \frac{1}{2^2} =$

$\frac{1}{4} + \frac{1}{4} = \frac{1}{2}$

16. $(5a^2b)^{-2}(5a^4b^{-2})^3$

$\frac{1}{(5a^2b)^2} \cdot (5a^4b^{-2})^3 =$

$\frac{125a^{12}b^{-6}}{25a^4b^2} = \frac{125a^{12} \cdot a^{-4}}{25b^2 \cdot b^6} =$

$\frac{125a^8}{25b^8} = \frac{5a^8}{b^8}$

17. $(6x^0y^{-1})^{-3}(2x^{-5}y^0)^{-2}$

$(6^{-3}y^3)(2^{-2}x^{10}) =$

$\frac{1}{6^3} \cdot y^3 \cdot \frac{1}{2^2} \cdot x^{10} =$

$\frac{1}{216} \cdot \frac{1}{4} \cdot x^{10}y^3 = \frac{1}{864}x^{10}y^3$

18. 57 $\left(\frac{5x^{-3}y}{3x^2y^{-2}}\right)^3 = \frac{5^3x^{-9}y^3}{3^3x^6y^{-6}} =$

$\frac{125y^3 \cdot y^6}{9x^9 \cdot x^6} = \frac{125y^9}{9x^{15}}$

27

19. 69 $\left(\frac{3x^3}{y^{-2}}\right)^{-2} \cdot \left(\frac{4x^5}{y^2}\right)^2$

$\frac{3^{-2}x^{-6}}{y^4} \cdot \frac{4^2x^{10}}{y^4} =$

$\frac{16x^4}{9y^8}$

20. $\frac{(x + 3y)^2}{(x + 3y)^{-3}}$

$(x+3y)^2 \cdot (x+3y)^3 =$

$(x+3y)^5$

j 11. $\left(\frac{a}{b}\right)^{-1}$ $\frac{b}{a}$ (k) $\frac{1}{a}$

b 12. $a^{-x}b^{-y}$ (l) $\frac{1}{a^y}$

Simplify each of the following. Remember to write your answers with positive exponents only.

13. $(4xy^3)(8x^3y^4)^2$

$(4xy^3)(64x^6y^8)$

$256x^7y^{11}$

14. $\left(\frac{2}{5}\right)^{-3} = \left(\frac{5}{2}\right)^3 = \frac{125}{8}$

15. / 23 $4^{-1} + 2^{-2}$

$\frac{1}{4} + \frac{1}{4} = \frac{2}{4} = \frac{1}{2}$

16. $(5a^2b)^{-2}(5a^4b^{-2})^3$

$\frac{1}{25a^4b^2} \cdot \frac{125a^{12}}{b^6} =$

$\frac{5a^8}{b^8}$

17. $(6x^0y^{-1})^{-3}(2x^{-5}y^0)^{-2}$

$(6^{-3}y^3)(2^{-2}x^{10})$

$\frac{1}{6^3} \cdot y^3 \cdot \frac{1}{2^2} \cdot x^{10}$

$\frac{y^3}{216} \cdot \frac{x^{10}}{4} = \frac{y^3x^{10}}{864}$

18. / 57 $\left(\frac{5x^{-3}y}{3x^2y^{-2}}\right)^3$

$\frac{125y^3y^6}{27x^6x^9} = \frac{125y^9}{27x^{15}}$

19. / 69 $\left(\frac{3x^3}{y^{-2}}\right)^{-2} \cdot \left(\frac{4x^5}{y^2}\right)^2$

$\frac{1}{9x^6y^4} \cdot \frac{16x^{10}}{y^4} =$

$\frac{16x^{10-6}}{9y^8}$

$\frac{16x^4}{9y^8}$

20. $\frac{(x + 3y)^2}{(x + 3y)^{-3}}$

$(x+3y)^2 \cdot (x+3y)^3 =$

$(x+3y)^5$

21. $(x + y)(x^{-1} + y^{-1})$

$(x+y)(\frac{1}{x}+\frac{1}{y}) =$

$\frac{x+y}{x} + \frac{x+y}{y} =$

$\frac{y(x+y)}{xy} + \frac{x(x+y)}{xy} =$

$\frac{xy+y^2}{xy} + \frac{x^2+xy}{xy} = \frac{x^2+2xy+y^2}{xy} =$

$\boxed{\frac{(x+y)^2}{xy}}$

22. 75 $\left[\left(\frac{2x^2}{3y^3}\right)^{-1}\right]^{-3} = \left[\frac{2^{-1}x^{-2}}{3^{-1}y^{-3}}\right]^{-3} =$

$\frac{2^3x^6}{3^3y^9} = \boxed{\frac{8x^6}{27y^9}}$

4·2 Reducing Rational Expressions to Lowest Terms

1. Any expression that can be written as the quotient of two polynomials, where the denominator is not equal to zero, is called a rational expression.

2. Examples showing method of reducing rational expressions to lowest terms:

   (a) Reduce $\frac{15x^5y^9}{6xy^3}$ to lowest terms.

   $15x^5y^9 = 3 \cdot 5 \cdot x^5 \cdot y^9$ (1) Factor the numerator.

   $6xy^3 = 3 \cdot 2 \cdot x \cdot y^3$ (2) Factor the denominator.

   Therefore the GCF of $15x^5y^9$ and $6xy^3$ is $3xy^3$.

   Note that when finding the GCF of variable factors such as y^3 and y^9, the factor with the smaller exponent is used in the GCF.

   $$\frac{15x^5y^9}{6xy^3} = \frac{3xy^3 \cdot 5x^4y^6}{3xy^3 \cdot 2} = \boxed{\frac{5x^4y^6}{2}}$$

21. $(x + y)(x^{-1} + y^{-1})$

$(x+y) \cdot \frac{1}{x+y}$

$\frac{x+y}{x} + \frac{x+y}{y}$

$\frac{y(x+y)}{x} + \frac{x(x+y)}{y}$

$\frac{xy+y^2}{xy} + \frac{x^2+xy}{xy} = \frac{x^2+2xy+y^2}{xy} = \frac{(x+y)^2}{xy}$

22. 75 $\left[\left(\frac{2x^2}{3y^3}\right)^{-1}\right]^{-3}$

$\left(\frac{2^{-1}x^{-2}}{3^{-1}y^{-3}}\right)^{-3} =$

$\frac{2^3x^6}{3^3y^9} = \frac{8x^6}{27y^9}$

4·2 Reducing Rational Expressions to Lowest Terms

1. Any expression that can be written as the quotient of two polynomials, where the denominator is not equal to zero, is called a rational expression.

2. Examples showing method of reducing rational expressions to lowest terms:

   (a) Reduce $\frac{15x^5y^9}{6xy^3}$ to lowest terms.

   $15x^5y^9 = 3 \cdot 5 \cdot x^5 \cdot y^9$ (1) Factor the numerator.

   $6xy^3 = 3 \cdot 2 \cdot x \cdot y^3$ (2) Factor the denominator.

   Therefore the GCF of $15x^5y^9$ and $6xy^3$ is $3xy^3$.

   Note that when finding the GCF of variable factors such as y^3 and y^9, the factor with the smaller exponent is used in the GCF.

   $$\frac{15x^5y^9}{6xy^3} = \frac{3xy^3 \cdot 5x^4y^6}{3xy^3 \cdot 2} = \frac{5x^4y^6}{2}$$

(b) Reduce $\dfrac{x^2 + 5x + 4}{x^2 - 16}$ to lowest terms.

$$\frac{x^2 + 5x + 4}{x^2 - 16} = \frac{(x + 4)(x + 1)}{(x + 4)(x - 4)}$$

(1) Factor numerator.

(2) Factor denominator.

$$= \boxed{\frac{x + 1}{x - 4}}$$

(3) Since $(x + 4)$ is the GCF, it can be divided out.

(c) Reduce $\dfrac{x^2 + 4x - 21}{9 - x^2}$ to lowest terms.

$$\frac{x^2 + 4x - 21}{9 - x^2} = \frac{(x + 7)(x - 3)}{(3 + x)(3 - x)} = \frac{(x + 7)(x - 3)}{(3 + x)[(-1)(-3 + x)]}$$

$$= \frac{(x + 7)(x - 3)}{(3 + x)(-1)(x - 3)} = \frac{x + 7}{(3 + x)(-1)} = \boxed{-\left(\frac{x + 7}{3 + x}\right)}$$

Reduce the following rational expressions to lowest terms.

3. $\dfrac{12x^5y^3}{22x^2y^8} = \dfrac{2x^2y^3 \cdot 6x^3}{2x^2y^3 \cdot 11y^5} = \boxed{\dfrac{6x^3}{11y^5}}$

$12x^5y^3 = 2 \cdot 2 \cdot 3 \cdot x^5 \cdot y^3$
$22x^2y^8 = 2 \cdot 11 \cdot x^2 \cdot y^8$

GCF $= 2x^2y^3$

4. $\dfrac{8x^3y^4}{2x^2y} = \dfrac{2x^2y \cdot 4xy^3}{2x^2y \cdot 1} = \boxed{4xy^3}$

$8x^3y^4 = 2 \cdot 2 \cdot 2 \cdot x^3 \cdot y^4$
$2x^2y = 2 \cdot x^2 \cdot y$

GCF $= 2x^3y$

5. 9 $\dfrac{-6xy^3z^2}{24x^3y^7z^8} = \dfrac{6xy3z^2 \cdot -1}{6xy^3z^2 \cdot 4x^2y^4z^6} = \boxed{\dfrac{-1}{4x^2y^4z^6}}$

GCF $= 6xy^3z^2$

(b) Reduce $\frac{x^2 + 5x + 4}{x^2 - 16}$ to lowest terms.

$$\frac{x^2 + 5x + 4}{x^2 - 16} = \frac{(x + 4)(x + 1)}{(x + 4)(x - 4)}$$

(1) Factor numerator.

(2) Factor denominator.

$$= \frac{x + 1}{x - 4}$$

(3) Since $(x + 4)$ is the GCF, it can be divided out.

(c) Reduce $\frac{x^2 + 4x - 21}{9 - x^2}$ to lowest terms.

$$\frac{x^2 + 4x - 21}{9 - x^2} = \frac{(x + 7)(x - 3)}{(3 + x)(3 - x)} = \frac{(x + 7)(x - 3)}{(3 + x)[(-1)(-3 + x)]}$$

$$= \frac{(x + 7)(x - 3)}{(3 + x)(-1)(x - 3)} = \frac{x + 7}{(3 + x)(-1)} = -\left(\frac{x + 7}{3 + x}\right)$$

Reduce the following rational expressions to lowest terms.

3. $\frac{12x^5y^3}{22x^2y^8}$ $= \frac{6x^3}{11y^5}$

4. $\frac{8x^3y^4}{2x^2y}$ $= 4xy^3$

5. $\frac{-6xy^3z^2}{24x^3y^7z^8}$ $= \frac{1}{-4x^2y^4z^6}$

6. $\dfrac{125a^3x^3(s-r)^3}{275a^4(s-r)^2} = \dfrac{25a^3(5-r)^2 \cdot 5x^3(s-r)}{25a^3(5-r)^2 \cdot 11a} = \boxed{\dfrac{5x^3(s-r)}{11a}}$

GCF $= 25a^3(s-r)^2$

7. $\dfrac{2x^2 - x - 15}{x^2 - 9} = \dfrac{(x-3)(2x+5)}{(x+3)(x-3)} = \boxed{\dfrac{2x+5}{x+3}}$

GCF $= (x-3)$

8. $\dfrac{2x^2 - x - 15}{9 - x^2} = \dfrac{(2x+5)(x-3)}{(3-x)(3+x)} - \dfrac{(2x+5)(3-x)}{(3-x)(3+x)} = \boxed{\dfrac{-(2x+5)}{3+x}}$

GCF $= (3-x)$

9. $\dfrac{-(a+b)}{-a-b} = \dfrac{-(a+b)}{-(a+b)} = \boxed{1}$

GCF $= -(a+b)$

10. [27] $\dfrac{3x^2 + 4x}{6x^2 + 5x - 4} = \dfrac{x(3x+4)}{(2x-1)(3x+4)} = \boxed{\dfrac{x}{2x-1}}$

GCF $= (3x+4)$

11. [33] $\dfrac{x^3 + 8}{3x^2 + 2x - 8} = \dfrac{(x+2)(x^2-2x+4)}{(x+2)(3x-4)} = \boxed{\dfrac{x^2-2x+4}{3x-4}}$

GCF $= (x+2)$

12. [39] $\dfrac{6ax - 8x - 3ay + 4y}{ay + 2y - 2ax - 4x} = \dfrac{(6ax-8x)-(3ay-4y)}{(ay+2y)-(2ax+4x)} = \dfrac{2x(3a-4)-y(3a-4)}{y(a+2)-2x(a+2)} =$

$\dfrac{(3a-4)(2x-y)}{(a+2)(y-2x)} = \dfrac{-(3a-4)(y-2x)}{(a+2)(y-2x)} = \boxed{\dfrac{-(3a-4)}{a+2}}$

GCF $= (y-2x)$

13. $\dfrac{x^2 + x + y - y^2}{1 - (x-y)^2} = \dfrac{(x^2-y^2)+(x+y)}{[1+(x-y)][1-(x-y)]} = \dfrac{(x+y)(x-y)+(x+y)}{[1+(x-y)][1-(x-y)]} =$

GCF $= [1+(x-y)]$ $\qquad \dfrac{(x+y)[(x-y)+1]}{[1+(x-y)][1-(x-y)]} = \dfrac{x+y}{1-(x-y)} = \boxed{\dfrac{x+y}{1-x+y}}$

14. $\dfrac{a^2 - 4}{8 - a^3} = \dfrac{-(4-a^2)}{8-a^3} = \dfrac{-(2-a)(2+a)}{(2-a)(4+2a+a^2)} = \boxed{\dfrac{-(2+a)}{a^2+2a+4}}$

GCF $= (2-a)$

6. $\dfrac{125a^3x^3(s - r)^3}{275a^4(s - r)^2}$ $\qquad \dfrac{5(s-r)}{11a}$

7. $\dfrac{2x^2 - x - 15}{x^2 - 9} = \dfrac{(2x+5)(x-3)}{(x-3)(x+3)} = \dfrac{2x+5}{x+3}$

8. $\dfrac{2x^2 - x - 15}{9 - x^2}$ $\quad \dfrac{(2x+5)(x-3)}{(3-x)(3+x)} = -\dfrac{2x+5}{3+x}$

9. $\dfrac{-(a + b)}{-a - b} = \dfrac{-a-b}{-a-b} = 1$

10. 27 $\dfrac{3x^2 + 4x}{6x^2 + 5x - 4} = \dfrac{x(3x+4)}{(2x-1)(3x+4)} = \dfrac{x}{2x-1}$

11. 33 $\dfrac{x^3 + 8}{3x^2 + 2x - 8} = \dfrac{x+2(x^2-2x+4)}{(3x-4)(x+2)} = \dfrac{x^2-2x+4}{3x-4}$

12. 39 $\dfrac{6ax - 8x - 3ay + 4y}{ay + 2y - 2ax - 4x} = \dfrac{2x(3a-4) - y(3a+4)}{y(a+2) - 2x(a-2)}$

$= \dfrac{(2x-y)(3a-4)}{(y-2x)(a+2)} = -\dfrac{3a-4}{a+2}$

? 13. $\dfrac{x^2 + x + y - y^2}{1 - (x - y)^2} = \dfrac{x^2-y^2+x+y}{1-(x-y)^2} = \dfrac{(x-y)(x+y)+x+y}{[1+(x-y)][1-(x-y)]}$

14. $\dfrac{a^2 - 4}{8 - a^3} = \dfrac{(a-2)(a+2)}{2+a(4-2a+a^2)} = \dfrac{a-2}{a^2-2a+4}$

15. $$\frac{a^2 - 2ax + x^2}{a^3 - a^2x - ax^2 + x^3} = \frac{(a-x)(a-x)}{a^2(a-x)-x^2(a-x)} = \frac{(a-x)(a-x)}{(a-x)(a^2-x^2)} =$$

GCF $=(a-x)(a-x)$ $$\frac{(a-x)(a-x)}{(a-x)(a-x)(a+x)} = \boxed{\frac{1}{a+x}}$$

4·3 Multiplying and Dividing Rational Expressions

1. Examples showing methods of (a) multiplying and (b) dividing rational expressions.

(a) Multiply: $\frac{5x^3y^4}{6x^2y^5} \cdot \frac{15xy^2}{40x^3y}$

$$= \frac{3 \cdot 5 \cdot 5 \cdot x^4 \cdot y^6}{2 \cdot 2 \cdot 2 \cdot 2 \cdot 3 \cdot 5 \cdot x^5y^6}$$ (1) Factor.

$$= \frac{(15x^4y^6)5}{(15x^4y^6)16x}$$ (2) Find the GCF.

$$= \boxed{\frac{5}{16x}}$$ (3) Divide out GCF.

(b) Divide: $\frac{(x^2 + 3x + 2)}{(x^2 - 9)} \div \frac{(x^2 - 4)}{(x^2 + 5x + 6)}$

$$= \frac{(x^2 + 3x + 2)}{(x^2 - 9)} \cdot \frac{(x^2 + 5x + 6)}{(x^2 - 4)}$$ (1) Invert and multiply.

$$= \frac{(x^2 + 3x + 2)(x^2 + 5x + 6)}{(x^2 - 9)(x^2 - 4)}$$

$$= \frac{(x+2)(x + 1)(x+3)(x + 2)}{(x + 3)(x - 3)(x+2)(x-2)}$$ (2) Factor and divide out common factors.

$$= \boxed{\frac{(x + 1)(x + 2)}{(x - 3)(x - 2)}}$$ (3) Leave final result in factored form.

15. $\dfrac{a^2 - 2ax + x^2}{a^3 - a^2x - ax^2 + x^3}$

4·3 Multiplying and Dividing Rational Expressions

1. Examples showing methods of (a) multiplying and (b) dividing rational expressions.

(a) Multiply: $\dfrac{5x^3y^4}{6x^2y^5} \cdot \dfrac{15xy^2}{40x^3y}$

$$= \frac{3 \cdot 5 \cdot 5 \cdot x^4 \cdot y^6}{2 \cdot 2 \cdot 2 \cdot 2 \cdot 3 \cdot 5 \cdot x^5y^6}$$ (1) Factor.

$$= \frac{(\qquad\qquad)5}{(\qquad\qquad)16x}$$ (2) Find the GCF.

$$= \frac{5}{16x}$$ (3) Divide out GCF.

(b) Divide: $\dfrac{(x^2 + 3x + 2)}{(x^2 - 9)} \div \dfrac{(x^2 - 4)}{(x^2 + 5x + 6)}$

$$= \frac{(x^2 + 3x + 2)}{(x^2 - 9)} \cdot \frac{(x^2 + 5x + 6)}{(x^2 - 4)}$$ (1) Invert and multiply.

$$= \frac{(x^2 + 3x + 2)(x^2 + 5x + 6)}{(x^2 - 9)(x^2 - 4)}$$

$$= \frac{(\qquad)(x + 1)(\qquad)(x + 2)}{(x + 3)(x - 3)(\qquad)(\qquad)}$$ (2) Factor and divide out common factors.

$$= \frac{(x + 1)(x + 2)}{(x - 3)(x - 2)}$$ (3) Leave final result in factored form.

Perform the indicated operations and reduce the answers to lowest terms.

2. $\frac{8m^2}{x^2} \cdot \frac{5mx}{4} \cdot \frac{2x}{m^3} = \frac{80m^3x^2}{4m^3x^2} = \frac{4m^3x^2 \cdot 20}{4m^3x^2} = \boxed{20}$

GCF $= 4m^3x^2$

3. $\frac{(m+n)}{(m-n)} \cdot \frac{(n^2-m^2)}{(m+n)^2} = \frac{(m+n)(n^2-m^2)}{(m-n)(m+n)^2} \overset{①}{=} \frac{(m+n)(n-m)(n+m)}{(m-n)(m+n)(m+n)} =$

① GCF $= (m+n)(m+n)$

② GCF $= (m-n)$

$\frac{n-m}{m-n} \overset{②}{=} \frac{-(m-n)}{(m-n)} = \boxed{-1}$

4. $\frac{2a^2-13a+15}{4a^2-9} \cdot \frac{2a+1}{2a-1} \div \frac{a-5}{2a-1} = \frac{(2a-3)(a-5)(2a+1)}{(2a-3)(2a+3)(2a-1)} \cdot \frac{(2a-1)}{(a-5)} =$

$\frac{(2a-3)(a-5)(2a+1)(2a-1)}{(2a-3)(2a+3)(2a-1)(a-5)} = \boxed{\frac{2a+1}{2a+3}}$

GCF $= (2a-3)(a-5)(2a-1)$

$x^2-6x-45$

$x^2-3x+9x-45$

5. $\frac{x^2+14x-15}{x^2+4x-5} \div \frac{x^2+12x-45}{x^2+6x-27} = \frac{(x+15)(x-1)}{(x+5)(x-1)} \cdot \frac{(x+9)(x-3)}{(x+15)(x-3)} =$ ✓ wrong

$\boxed{\frac{x+9}{x+5}}$

6. $\frac{x^4+x}{x^2+x+1} \div \frac{x^2-x+1}{x^4-x} = \frac{x(x^3+1)}{x^2+x+1} \cdot \frac{x(x^3-1)}{x^2-x+1} =$

$\frac{x(x+1)(x^2-x+1)}{x^2+x+1} \cdot \frac{x(x-1)(x^2+x+1)}{x^2-x+1} =$

$\boxed{x^2(x+1)(x-1)}$ $\quad x^3+x^2+x^3-x^2$

$2x^3$

7. 61 $\quad \frac{(2x^3-10x^2-x+5)}{(x^3-5x^2+3x-15)} \div \frac{(2x^3+6x^2-x-3)}{(x^3+3x^2+3x+9)} =$

$\frac{2x(x-5)-(x-5)}{x^2(x-5)+3(x-5)} \cdot \frac{x^2(x+3)+3(x+3)}{2x^2(x+3)-(x+3)} = \frac{(x-5)(2x-1)}{(x-5)(x^2+3)} \cdot \frac{(x+3)(x^2+3)}{(x+3)(2x-1)} = \boxed{1}$

8. 51 $\quad \frac{(x^2+2x-15)}{(2x^2-5x-3)} \div (x^2+x-20) =$

$\frac{(x+5)(x-3)}{(2x+1)(x-3)} \cdot \frac{1}{(x+5)(x-4)} = \boxed{\frac{1}{(2x+1)(x-4)}}$

Perform the indicated operations and reduce the answers to lowest terms.

2. $\dfrac{8m^2}{x^2} \cdot \dfrac{5mx}{4} \cdot \dfrac{2x}{m^3}$ $= \dfrac{8m^2\,5mx\,2x}{x^2\,4m^3} = \dfrac{80m^3x^2}{4m^3x^2} = 20$

= 20

3. $\dfrac{(m+n)}{(m-n)} \cdot \dfrac{(n^2-m^2)}{(m+n)^2}$ $\dfrac{n^2-m^2}{(m+n)(m-n)} = \dfrac{(-1)(+1)\,(n-m)(n+m)}{(m+n)(m-n)} = -1$

$m+n$

4. $\dfrac{2a^2-13a+15}{4a^2-9} \cdot \dfrac{2a+1}{2a-1} \div \dfrac{a-5}{2a-1}$ $= \dfrac{(2a-3)(a-5)(2a+1)}{(2a-3)(2a+3)(2a-1)} \cdot \dfrac{(2a-1)}{a-5}$

$\dfrac{2a+1}{2a+3}$

9:5
12:4

5. $\dfrac{x^2+14x-15}{x^2+4x-5} \div \dfrac{x^2+12x-45}{x^2+6x-27}$ $= \dfrac{(x-1)(x+15)}{(x+5)(x-1)} = \dfrac{x+15}{x+5} \cdot \dfrac{(x\quad)(\quad)}{(x-3)(x+9)}$

Wrong answer in book

6. $\dfrac{x^4+x}{x^2+x+1} \div \dfrac{x^2-x+1}{x^4-x}$

$\dfrac{x(x+1)[x^2-x+1]}{[x^2+x+1]} \cdot \dfrac{x(x-1)[x^2+x+1]}{(x^2-x+1)} = \dfrac{x(x+1)x(x-1)}{x^2(x+1)(x-1)}$

7. (61) $\dfrac{(2x^3-10x^2-x+5)}{(x^3-5x^2+3x-15)} \div \dfrac{(2x^3+6x^2-x-3)}{(x^3+3x^2+3x+9)}$

8. (51) $\dfrac{(x^2+2x-15)}{(2x^2-5x-3)} \div (x^2+x-20)$

4·4 Adding and Subtracting Rational Expressions

1. (a) Find the Lowest Common Denominator (LCD) of $x^2 - 5x$ and $x^3 - 125$.

$$x^2 - 5x = x \;(x - 5)$$
$$x^3 - 125 = (x - 5)(x^2 + 5x + 25)$$
$$\text{LCD} = x\;(x - 5)(x^2 + 5x + 25)$$

(b) Add: $\dfrac{3}{x^2 - 5x} + \dfrac{2x}{x^3 - 125}$

$= \dfrac{3}{x(x - 5)} + \dfrac{2x}{(x - 5)(x^2 + 5x + 25)}$ (1) Factor denominator.

$= \dfrac{3}{x(x - 5)} + \dfrac{2x}{(x - 5)(x^2 + 5x + 25)}$ (2) Multiply by factor needed to result in the LCD.

$= \dfrac{3(x^2 + 5x + 25)}{x(x - 5)(x^2 + 5x + 25)} + \dfrac{2x(x)}{(x - 5)(x^2 + 5x + 25)(x)}$

$= \dfrac{3x^2 + 15x + 75 + 2x^2}{x(x - 5)(x^2 + 5x + 25)}$ (3) Write numerator over LCD.

$= \dfrac{5x^2 + 15x + 75}{x(x - 5)(x^2 + 5x + 25)}$ (4) Combine like terms in numerator.

$= \dfrac{5(x^2 + 3x + 15\quad)}{x(x - 5)(x^2 + 5x + 25)}$ (5) Reduce to lowest terms.

Perform the indicated operations and reduce answers to lowest terms.

2. $\dfrac{4x + 13}{1 - x^2} - \dfrac{2}{x + x^2} =$

$\dfrac{4x+13}{(1+x)(1-x)} - \dfrac{2}{x(1+x)} =$

$\dfrac{(4x+13)(x)}{(1+x)(1-x)(x)} - \dfrac{2(1-x)}{x(1+x)(1-x)} =$

$\dfrac{4x^2+13x-2+2x}{(1+x)(1-x)(x)} = \boxed{\dfrac{4x^2+15x-2}{x(x+1)(1-x)}}$

3. $\dfrac{4}{3x^2} - \dfrac{5}{2x^2} + \dfrac{6}{5x^3} =$

$\dfrac{4(10x)}{3x^2(10x)} - \dfrac{5(15x)}{2x^2(15x)} + \dfrac{6(6)}{5x^3(6)} =$

$\dfrac{40x}{30x^3} - \dfrac{75x}{30x^3} + \dfrac{36}{30x^3} =$

$\dfrac{40x-75x+36}{30x^3} = \boxed{\dfrac{-35x+36}{30x^3}}$

4·4 Adding and Subtracting Rational Expressions

1. (a) Find the Lowest Common Denominator (LCD) of $x^2 - 5x$ and $x^3 - 125$.

$$x^2 - 5x = x \quad (x - 5)$$

$$x^3 - 125 = \quad (x - 5) \quad (x^2 + 5x + 25)$$

$$\text{LCD} = x \quad (x - 5) \quad (x^2 + 5x + 25)$$

(b) Add: $\dfrac{3}{x^2 - 5x} + \dfrac{2x}{x^3 - 125}$

$$= \frac{3}{x(x - 5)} + \frac{2x}{(x - 5)(x^2 + 5x + 25)}$$

(1) Factor denominator.

$$= \frac{3}{x(x - 5)} + \frac{2x}{(x - 5)(x^2 + 5x + 25)}$$

(2) Multiply by factor needed to result in the LCD.

$$= \frac{3(x^2 + 5x + 25)}{x(x - 5)(x^2 + 5x + 25)} + \frac{2x(x)}{(x - 5)(x^2 + 5x + 25)(x)}$$

$$= \frac{3x^2 + 15x + 75 + 2x^2}{x(x - 5)(x^2 + 5x + 25)}$$

(3) Write numerator over LCD.

$$= \frac{\qquad\qquad}{x(x - 5)(x^2 + 5x + 25)}$$

(4) Combine like terms in numerator.

$$= \frac{5(\qquad\qquad)}{x(x - 5)(x^2 + 5x + 25)}$$

(5) Reduce to lowest terms.

Perform the indicated operations and reduce answers to lowest terms.

2. $\dfrac{4x + 13}{1 - x^2} - \dfrac{2}{x + x^2}$

3. $\dfrac{4}{3x^2} - \dfrac{5}{2x^2} + \dfrac{6}{5x^3}$

4. $\frac{x + 3}{4x^2 - 1} + \frac{x - 2}{1 - 4x^2} =$

$\frac{(x+3)}{4x^2-1} + \frac{-(x-2)}{4x^2-1} =$

$\frac{x+3-x+2}{(2x+1)(2x-1)} =$

$\boxed{\frac{5}{(2x+1)(2x-1)}}$

5. | 45 $\frac{3}{x^2 - 5x} + \frac{7}{25 - x^2} =$

$\frac{3}{x^2-5x} + \frac{-7}{x^2-25} =$

$\frac{3}{x(x-5)} + \frac{-7}{(x+5)(x-5)} =$

$\frac{3(x+5)}{x(x-5)(x+5)} + \frac{-7(x)}{(x+5)(x-5)(x)} =$

$\frac{3x+15-7x}{x(x-5)(x+5)} = \boxed{\frac{-4x+15}{x(x-5)(x+5)}}$

6. $\frac{3x + y}{x - y} + \frac{1}{4(x - y)^2} + 1 =$

$\frac{(3x+y)(4)(x-y)}{(x-y)(4)(x-y)} + \frac{1}{4(x-y)(x-y)} + \frac{4(x-y)(x-y)}{4(x-y)(x-y)} =$

$\frac{12x^2-8xy-4y^2+1+4x^2-8xy+4y^2}{4(x-y)(x-y)} =$

$\boxed{\frac{16x^2-16xy+1}{4(x-y)(x-y)}}$

7. $\frac{x^2 - 4}{x^2 + 3x + 2} - \frac{x - 2}{x + 2} =$

$\frac{x^2-4}{(x+2)(x+1)} - \frac{x-2}{x+2} =$

$\frac{x^2-4}{(x+2)(x+1)} - \frac{(x-2)(x+1)}{(x+2)(x+1)} =$

$\frac{x^2-4-(x^2-x-2)}{(x+2)(x+1)} =$

$\frac{x^2-4-x^2+x+2}{(x+2)(x+1)} = \boxed{\frac{x-2}{(x+2)(x+1)}}$

8. | 51 $\frac{4x}{3x^2 - 5x - 2} - \frac{1}{3x^2 + 13x + 4} =$

$\frac{4x}{(3x+1)(x-2)} - \frac{1}{(3x+1)(x+4)} =$

$\frac{4x(x+4)}{(3x+1)(x-2)(x+4)} - \frac{1(x-2)}{(3x+1)(x+4)(x-2)} =$

$\frac{4x^2+16x-x+2}{(3x+1)(x-2)(x+4)} =$

$\boxed{\frac{4x^2+15x+2}{(3x+1)(x-2)(x+4)}}$

9. $\frac{3x - 4}{2x^2 - 3x - 5} - \frac{4x + 3}{x + 1} =$

$\frac{(3x-4)}{(2x-5)(x+1)} - \frac{(4x+3)(2x-5)}{(x+1)(2x-5)} =$

$\frac{3x-4-8x^2+14x+15}{(2x-5)(x+1)} =$

$\boxed{\frac{-8x^2+17x+11}{(2x-5)(x+1)}}$

4. $\dfrac{x + 3}{4x^2 - 1} + \dfrac{x - 2}{1 - 4x^2}$

5. [45] $\dfrac{3}{x^2 - 5x} + \dfrac{7}{25 - x^2}$

6. $\dfrac{3x + y}{x - y} + \dfrac{1}{4(x - y)^2} + 1$

7. $\dfrac{x^2 - 4}{x^2 + 3x + 2} - \dfrac{x - 2}{x + 2}$

8. [51] $\dfrac{4x}{3x^2 - 5x - 2} - \dfrac{1}{3x^2 + 13x + 4}$

9. $\dfrac{3x - 4}{2x^2 - 3x - 5} - \dfrac{4x + 3}{x + 1}$

10. 83

$$\frac{3x+1}{x^2+x-20} - \frac{7}{x^2+9x+20} - \frac{x-3}{x^2-16} =$$

$$\frac{3x+1}{(x+5)(x-4)} - \frac{7}{(x+5)(x+4)} - \frac{x-3}{(x+4)(x-4)} =$$

$$\frac{(3x+1)\quad(x+4)}{(x+5)(x-4)(x+4)} - \frac{7\quad(x-4)}{(x+5)(x+4)(x-4)} - \frac{(x-3)\quad(x+5)}{(x+4)(x-4)(x+5)} =$$

$$\frac{3x^2+13x+4-7x+28-x^2-2x+15}{(x+5)(x-4)(x+4)} =$$

$$\boxed{\frac{2x^2+4x+47}{(x+5)(x-4)(x+4)}}$$

4·5 Complex Fractions

Rewrite each complex fraction in the form $a \div b$, then invert and multiply to simplify.

1. $$\frac{\frac{3}{5}}{\frac{27}{35}} =$$

$$\frac{3}{5} \div \frac{27}{35} =$$

$$\frac{3}{5} \cdot \frac{35}{27} = \boxed{\frac{7}{9}}$$

2. $$\frac{\frac{4}{x} + \frac{3}{y}}{\frac{8}{xy^2}} =$$

$$\left(\frac{4}{x} + \frac{3}{y}\right) \cdot \frac{xy^2}{8} =$$

$$\frac{4xy^2}{8x} + \frac{3xy^2}{8y} = \frac{4xy^3}{8xy} + \frac{3x^2y^2}{8xy} =$$

$$\boxed{\frac{4xy^3+3x^2y^2}{8xy}}$$

10. | 83

$$\frac{3x + 1}{x^2 + x - 20} - \frac{7}{x^2 + 9x + 20} - \frac{x - 3}{x^2 - 16}$$

4·5 Complex Fractions

Rewrite each complex fraction in the form $a \div b$, then invert and multiply to simplify.

1. $\dfrac{\frac{3}{5}}{\frac{27}{35}}$

$\frac{3}{5} \div \frac{27}{35}$

$\frac{3}{5} \cdot \frac{35}{27}$

$= \frac{7}{9}$

2. $\dfrac{\frac{4}{x} + \frac{3}{y}}{\frac{8}{xy^2}}$

$\frac{4}{x} + \frac{3}{y} \div \frac{8}{xy^2}$

$\left(\frac{4}{x} + \frac{3}{y}\right) \cdot \frac{xy^2}{8}$

$(y)\ \frac{4xy^2}{8x} + \frac{3xy^2}{8y}\ (x)$

$= \frac{4xy^3}{8xy} + \frac{3x^2y^2}{8xy}$

$= \frac{4xy^3 + 3x^2y^2}{8xy}$

3. 17

$$\frac{\frac{x+7}{4x^2}}{\frac{x-1}{8x}} =$$

$$\frac{x+7}{4x^2} \cdot \frac{8x}{x-1} =$$

$$\boxed{\frac{2(x+7)}{x(x-1)}}$$

4.

$$\frac{\frac{2x-6}{5xy^3}}{\frac{12x^2y - 6xy}{9x^2y^2}} =$$

$$\frac{2(x-3)}{5xy^3} \cdot \frac{9x^2y^2}{6xy(2x-1)} =$$

$$\boxed{\frac{3(x-3)}{5y^2(2x-1)}}$$

5.

$$\frac{\frac{1-x}{1+x}}{\frac{x-1}{2x}} =$$

$$\frac{1-x}{1+x} \cdot \frac{2x}{x-1} =$$

$$\frac{-(x-1)}{(1+x)} \cdot \frac{2x}{(x-1)} =$$

$$\boxed{\frac{-2x}{x+1}}$$

6.

$$\frac{\frac{3}{x} - \frac{5}{y}}{15} =$$

$$\left(\frac{3 \cdot y}{x \cdot y} - \frac{5 \cdot x}{y \cdot x}\right) \cdot \frac{1}{15} =$$

$$\left(\frac{3y - 5x}{xy}\right) \cdot \frac{1}{15} =$$

$$\boxed{\frac{3y-5x}{15xy}}$$

7. 41

$$\frac{2x - y}{\frac{2}{y} - \frac{1}{x}} =$$

$$\frac{2x-y}{1} \div \left(\frac{2 \cdot x}{yx} - \frac{1y}{xy}\right) =$$

$$\frac{2x-y}{1} \div \frac{2x-y}{xy} =$$

$$\frac{2x-y}{1} \cdot \frac{xy}{2x-y} = \boxed{xy}$$

8. 75

$$\frac{5x^{-1} + 15x^{-2}}{2 + 6x^{-1}} = \frac{\frac{5}{x} + \frac{15}{x^2}}{2 + \frac{6}{x}} =$$

$$\left(\frac{5 \cdot x}{x \cdot x} + \frac{15}{x^2}\right) \div \left(\frac{2 \cdot x}{1 \cdot x} + \frac{6}{x}\right) =$$

$$\frac{5x+15}{x^2} \div \frac{2x+6}{x} =$$

$$\frac{5(x+3)}{x^2} \cdot \frac{x}{2(x+3)} = \boxed{\frac{5}{2x}}$$

Multiply both the numerator and denominator of each complex fraction by the lowest common denominator (LCD) of all fractions contained in the complex fraction.

Example:

$$\frac{\frac{4}{x} + \frac{3}{y}}{\frac{8}{xy^2}} \cdot \frac{xy^2}{xy^2} = \frac{\frac{4xy^2}{x} + \frac{3xy^2}{y}}{\frac{8xy^2}{xy^2}} = \frac{4y^2 + 3xy}{8}$$

3. 17 $\dfrac{\dfrac{x + 7}{4x^2}}{\dfrac{x - 1}{8x}}$ $\quad \dfrac{x+7}{4x^2} \div \dfrac{x-1}{8x}$

$\dfrac{(x+7)}{4x^2} \cdot \dfrac{8x}{x-1} = \dfrac{2(x+7)}{x(x-1)}$

4. $\dfrac{\dfrac{2x - 6}{5xy^3}}{\dfrac{12x^2y - 6xy}{9x^2y^2}}$ $\quad \dfrac{2(x-3)}{5xy^3} \cdot \dfrac{9x^2y^2}{6xy(2x-1)}$

$\dfrac{(x-3)3}{5y^2(2x-1)}$

5. $\dfrac{\dfrac{1 - x}{1 + x}}{\dfrac{x - 1}{2x}}$ $\quad \dfrac{1-x}{1+x} \cdot \dfrac{2x}{x-1}$

$\dfrac{-2x}{1+x}$

6. $\dfrac{\dfrac{3}{x} - \dfrac{5}{y}}{15}$ $\quad \left(\dfrac{3}{x} - \dfrac{5}{y}\right) \cdot \dfrac{1}{15}$

$\left(\dfrac{3y - 5x}{xy}\right) \cdot \dfrac{1}{15} = \dfrac{3y - 5x}{15xy}$

7. 41 $\dfrac{2x - y}{\dfrac{2}{y} - \dfrac{1}{x}}$ $\quad 2x - y \div \dfrac{2}{y} - \dfrac{1}{x}$

$2x - y \div \left(\dfrac{2x-y}{xy}\right)$

$\dfrac{2x-y}{1} \cdot \dfrac{xy}{2x-y}$

$= \dfrac{xy}{1} = xy$

8. 75 $\dfrac{5x^{-1} + 15x^{-2}}{2 + 6x^{-1}} = \dfrac{\dfrac{5}{x} + \dfrac{15}{x^2}}{2 + \dfrac{6}{x}}$

$\dfrac{x}{x}\left(\dfrac{5}{x} + \dfrac{15}{x^2}\right) \div \dfrac{x}{x}\left(\dfrac{2}{1} + \dfrac{6}{x}\right)$

$\left(\dfrac{5x}{x^2} + \dfrac{15}{x^2}\right) \div \left(\dfrac{2x}{x} + \dfrac{6}{x}\right) = \dfrac{(5x+15)}{x^2}\dfrac{(x)}{(2x+6)}$

$\dfrac{5x(x+3)}{2x^3+6x^2} = \dfrac{5x(x+3)}{2x^2(x+3)} = \dfrac{5x}{2x^2} = \dfrac{5}{2x}$

Multiply both the numerator and denominator of each complex fraction by the lowest common denominator (LCD) of all fractions contained in the complex fraction.

Example: $$\frac{\dfrac{4}{x} + \dfrac{3}{y}}{\dfrac{8}{xy^2}} \cdot \frac{xy^2}{xy^2} = \frac{\dfrac{4xy^2}{x} + \dfrac{3xy^2}{y}}{\dfrac{8xy^2}{xy^2}} = \frac{4y^2 + 3xy}{8}$$

$(xy^2)\dfrac{4}{x} + \dfrac{3}{y}(xy^2)$

$4y^2 + 3xy \div 8 = \dfrac{4y^2 + 3xy}{8}$

9. | 17

$$\frac{\frac{x+7}{4x^2}}{\frac{x-1}{8x}} \cdot \frac{8x^2}{8x^2} =$$

$$\frac{\frac{8x^2(x+7)}{4x^2}}{\frac{8x^2(x-1)}{8x}} = \boxed{\frac{2(x+7)}{x(x-1)}}$$

10.

$$\frac{\frac{2x-6}{5xy^3}}{\frac{12x^2y-6xy}{9x^2y^2}} \cdot \frac{45x^2y^3}{45x^2y^3} =$$

$$\frac{\frac{45x^2y^3(2)(x-3)}{5xy^3}}{\frac{45x^2y^3(6xy)(2x-1)}{9x^2y^2}} = \frac{9x(2)(x-3)}{5y(6xy)(2x-1)} =$$

$$\boxed{\frac{3(x-3)}{5y^2(2x-1)}}$$

11. | 41

$$\frac{2x-y}{\frac{2}{y}-\frac{1}{x}} \cdot \frac{xy}{xy} =$$

$$\frac{xy(2x-y)}{\frac{2xy}{y}-\frac{xy}{x}} = \frac{xy(2x-y)}{(2x-y)} = \boxed{xy}$$

12. | 75

$$\frac{5x^{-1}+15x^{-2}}{2+6x^{-1}} = \frac{\frac{5}{x}+\frac{15}{x^2}}{2+\frac{6}{x}} \cdot \frac{x^2}{x^2} =$$

$$\frac{\frac{5x^2}{x}+\frac{15x^2}{x^2}}{2x^2+\frac{6x^2}{x}} = \frac{5x+15}{2x^2+6x} =$$

$$\frac{5(x+3)}{2x(x+3)} = \boxed{\frac{5}{2x}}$$

---

4 · 6 Division of Polynomials

Use the property $\frac{a+b+c}{d} = \frac{a}{d}+\frac{b}{d}+\frac{c}{d}$ to simplify each expression.

Remember that this property is best used when the <u>divisor</u> is a <u>monomial</u>.

1. $$\frac{25xy^3+15x^2y-5x^4}{5x} = \frac{25xy^3}{5x}+\frac{15x^2y}{5x}-\frac{5x^4}{5x} = \boxed{5y^3+3xy-x^3}$$

2. | 7

$$\frac{10x^4y^2-16x^3y^3-2x^2y^4}{2x^2y^2} = \frac{10x^4y^2}{2x^2y^2}-\frac{16x^3y^3}{2x^2y^2}-\frac{2x^2y^4}{2x^2y^2} =$$

$$\boxed{5x^2-8xy-y^2}$$

9. $\dfrac{\dfrac{x + 7}{4x^2}}{\dfrac{x - 1}{8x}}$

10. $\dfrac{\dfrac{2x - 6}{5xy^3}}{\dfrac{12x^2y - 6xy}{9x^2y^2}}$

11. $\dfrac{2x - y}{\dfrac{2}{y} - \dfrac{1}{x}}$

12. $\dfrac{5x^{-1} + 15x^{-2}}{2 + 6x^{-1}}$

4·6 Division of Polynomials

Use the property $\dfrac{a + b + c}{d} = \dfrac{a}{d} + \dfrac{b}{d} + \dfrac{c}{d}$ to simplify each expression.

Remember that this property is best used when the divisor is a monomial.

1. $\dfrac{25xy^3 + 15x^2y - 5x^4}{5x}$

2. $\dfrac{10x^4y^2 - 16x^3y^3 - 2x^2y^4}{2x^2y^2}$

3. $\dfrac{-16a^{2m} - 6a^{6m}}{-2a^{m}} = \dfrac{-16a^{2m}}{-2a^{m}} - \dfrac{6a^{6m}}{-2a^{m}} = \boxed{8a^{m} + 3a^{5m}}$

4. $\dfrac{-120x^4y^3z^7 + 40x^2y^5z^4 - 30x^6y^2z^8}{-10x^2y^2z^4} = \dfrac{-120x^4y^3z^7}{-10x^2y^2z^4} + \dfrac{40x^2y^5z^4}{-10x^2y^2z^4} -$

$\dfrac{30x^6y^2z^8}{-10x^2y^2z^4} = \boxed{12x^2yz^3 + {}^{-}4y^3 + 3x^4z^4}$

5. | 13

$\dfrac{3x - 10x^4 - 72z^4}{8x^2yz} = \dfrac{3x}{8x^2yz} - \dfrac{10x^4}{8x^2yz} - \dfrac{72z^4}{8x^2yz} =$

$\boxed{\dfrac{3}{8xyz} - \dfrac{5x^2}{4yz} - \dfrac{9z^3}{x^2y}}$

Use the long-division method to write each expression in lowest terms.

Remember to insert a coefficient of zero for each missing power.

6. $(x^4 - 2x^2 + 1) \div (x^2 + 2x + 1)$

Hint: Rewrite as $(x^4 + 0x^3 - 2x^2 + 0x + 1) \div (x^2 + 2x + 1)$

$$
\begin{array}{r}
x^2 - 2x + 1 \\
x^2+2x+1 \overline{) x^4 + 0x^3 - 2x^2 + 0x + 1} \\
\underline{x^4 \mp 2x^3 \mp x^2} \\
-2x^3 - 3x^2 + 0x + 1 \\
\underline{\overset{(+)}{-}2x^3 \overset{(+)}{-}4x^2 \overset{(+)}{-}2x} \\
x^2 + 2x + 1 \\
\underline{-x^2 \mp 2x \mp 1} \\
0
\end{array}
$$

$\boxed{x^2 - 2x + 1}$

3. $\dfrac{-16a^{2m} - 6a^{6m}}{-2a^{m}}$

4. $\dfrac{-120x^4y^3z^7 + 40x^2y^5z^4 - 30x^6y^2z^8}{-10x^2y^2z^4}$

5. 13 $\dfrac{3x - 10x^4 - 72z^4}{8x^2yz}$

Use the long-division method to write each expression in lowest terms. Remember to insert a coefficient of zero for each missing power.

6. $(x^4 - 2x^2 + 1) \div (x^2 + 2x + 1)$

Hint: Rewrite as $(x^4 + 0x^3 - 2x^2 + 0x + 1) \div (x^2 + 2x + 1)$

7. $(x^2 - 7x + 16) \div (x - 5)$

$$\begin{array}{r} x - 2 + \frac{6}{x-5} \\ x-5 \overline{) x^2 - 7x + 16} \\ \underline{x^2 - 5x} \\ -2x + 16 \\ \underline{-2x + 10} \\ 6 \end{array}$$

$\boxed{x - 2 + \frac{6}{x-5}}$

8. (29) $\dfrac{40x^3 + 31x^2 + 16x + 8}{5x + 2}$

$$\begin{array}{r} 8x^2 + 3x + 2 + \frac{4}{5x+2} \\ 5x+2 \overline{) 40x^3 + 31x^2 + 16x + 8} \\ \underline{-40x^3 + {}^{-}16x^2} \\ 15x^2 + 16x + 8 \\ \underline{-15x^2 + {}^{-}6x} \\ 10x + 8 \\ \underline{-10x + {}^{-}4} \\ 4 \end{array}$$

$\boxed{8x^2 + 3x + 2 + \frac{4}{5x+2}}$

9. (57) $(2x^4 - 12x^3 + 27x - 3) \div (2x^2 - 4) =$

$$\begin{array}{r} x^2 - 6x + 2 + \frac{3x+5}{2x^2-4} \\ 2x^2-4 \overline{) 2x^4 - 12x^3 + 0x^2 + 27x - 3} \\ \underline{2x^4 \qquad - 4x^2} \\ -12x^3 + 4x^2 + 27x - 3 \\ \underline{-12x^3 \qquad + 24x} \\ 4x^2 + 3x - 3 \\ \underline{4x^2 \qquad - 8} \\ 3x + 5 \end{array}$$

$\boxed{x^2 - 6x + 2 + \frac{3x+5}{2x^2-4}}$

10. $\dfrac{8x^3 - 27}{2x - 3} =$

$$\begin{array}{r} 4x^2 + 6x + 9 \\ 2x-3 \overline{) 8x^3 + 0x^2 + 0x - 27} \\ \underline{8x^3 - 12x^2} \\ 12x^2 + 0x - 27 \\ \underline{12x^2 - 18x} \\ +18x - 27 \\ \underline{18x - 27} \\ 0 \end{array}$$

$\boxed{4x^2 + 6x + 9}$

7. $(x^2 - 7x + 16) \div (x - 5)$

8. 29 $\dfrac{40x^3 + 31x^2 + 16x + 8}{5x + 2}$

9. 57 $(2x^4 - 12x^3 + 27x - 3) \div (2x^2 - 4)$

10. $\dfrac{8x^3 - 27}{2x - 3}$

4 · 7 Synthetic Division

Divide using synthetic division.

1. $(2x^2 - 9) \div (x + 2)$

   $(2x^2 + 0x - 9) \div (x + 2)$ (1) Replace missing powers.

   (Line 1) $-2 \rfloor$ 2 0 -9

   (Line 2) (b) -4 8 (d)

   (Line 3) (a) 2 (c) -4 -1 (e)

   $2x - 4 + \frac{-1}{x+2}$ (f)

   (2) Fill in according to steps (a) through (f) given below.

   Steps:

   (a) 2 is brought down from Line 1 to Line 3.

   (b) -4, the product of -2 and 2, is written under the 0 on Line 2.

   (c) -4, the sum of zero and -4, is written on Line 3.

   (d) 8, the product of -2 and -4, is written under the -9.

   (e) -1, the sum of -9 and 8, is written on Line 3.

   (f) Rewrite the 2 -4 -1 found in Line 3 as $2x - 4 + \dfrac{-1}{x + 2}$.

   Therefore, $\dfrac{2x^2 - 9}{x + 2} = 2x - 4 + \dfrac{-1}{x + 2}$

2. (13) $\dfrac{4x^3 - 5x + 1}{x + 1/2}$

   $-\frac{1}{2}\rfloor$ 4 0 -5 1

   -2 1 2

   4 -2 -4 3

   $4x^2 - 2x - 4 + \dfrac{3}{x + \frac{1}{2}}$

3. $\dfrac{3x^3 + 2x^2 - 5}{x - 2}$

   $2\rfloor$ 3 2 0 -5

   6 16 32

   3 8 16 27

   $3x^2 + 8x + 16 + \dfrac{27}{x-2}$

4. $\dfrac{x^3 - 27}{x - 3}$

   $3\rfloor$ 1 0 0 -27

   3 9 27

   1 3 9 0

   $x^2 + 3x + 9$

5. (21) $\dfrac{16x^4 - 9x^2 + 4x - 2}{x + 3/4}$

   $-\frac{3}{4}\rfloor$ 16 0 -9 4 -2

   -12 9 0 -3

   16 -12 0 4 -5

   $16x^3 - 12x^2 + 4 - \dfrac{5}{x + 3/4}$

4·7 Synthetic Division

Divide using synthetic division.

1. $(2x^2 - 9) \div (x + 2)$

   $(2x^2 + 0x - 9) \div (x + 2)$ (1) Replace missing powers.

(Line 1)	-2	2	0	-9
(Line 2)				
(Line 3)				

   (2) Fill in according to steps (a) through (f) given below.

   Steps:

   (a) 2 is brought down from Line 1 to Line 3.

   (b) -4, the product of -2 and 2, is written under the 0 on Line 2.

   (c) -4, the sum of zero and -4, is written on Line 3.

   (d) 8, the product of -2 and -4, is written under the -9.

   (e) -1, the sum of -9 and 8, is written on Line 3.

   (f) Rewrite the 2 -4 -1 found in Line 3 as $2x - 4 + \dfrac{-1}{x + 2}$.

   Therefore, $\dfrac{2x^2 - 9}{x + 2} = 2x - 4 + \dfrac{-1}{x + 2}$

2. $\dfrac{4x^3 - 5x + 1}{x + 1/2}$

3. $\dfrac{3x^3 + 2x^2 - 5}{x - 2}$

4. $\dfrac{x^3 - 27}{x - 3}$

5. $\dfrac{16x^4 - 9x^2 + 4x - 2}{x + 3/4}$

4·8 Equations Involving Rational Expressions

1. A/an extraneous solution to a fractional equation is one that does not satisfy the equation because it results in division by zero.

2. To solve a fractional equation, multiply both sides of the equation by the LCD of the fractions in the equation.

Solve for the variable in each of the following equations.

3. $\frac{3}{2x} + \frac{1}{3x} = \frac{11}{12}$ [3]

$12x \cdot \left(\frac{3}{2x} + \frac{1}{3x}\right) = \left(\frac{11}{12}\right) \cdot 12x$

$18 + 4 = 11x$

$22 = 11x$

$\boxed{x = 2}$

4. $\frac{2x}{3} - \frac{x}{2} = \frac{1}{4}$

$12 \cdot \left(\frac{2x}{3} - \frac{x}{2}\right) = \left(\frac{1}{4}\right) \cdot 12$

$8x - 6x = 3$

$2x = 3$

$\boxed{x = \frac{3}{2}}$

5. $\frac{x}{x + 2} - \frac{4}{x + 1} = \frac{-2}{x + 2}$

$(x+2)(x+1)\left(\frac{x}{x+2} - \frac{4}{x+1}\right) = \left(\frac{-2}{x+2}\right)(x+2)(x+1)$

$x(x+1) - 4(x+2) = -2(x+1)$

$x^2 + x - 4x - 8 = -2x - 2$

$x^2 - x - 6 = 0$

$(x-3)(x+2) = 0$

$\boxed{x = 3}$ OR $x = -2$

Note: -2 is an extraneous root because it would yield division by zero if substituted into the original equation.

6. $\frac{y + 1}{y + 7} = \frac{y - 4}{y - 1}$

$(y-1)(y+7)\left(\frac{y+1}{y+7}\right) = \left(\frac{y-4}{y-1}\right)(y-1)(y+7)$

$(y-1)(y+1) = (y-4)(y+7)$

$y^2 - 1 = y^2 + 3y - 28$

$27 = 3y$

$\boxed{y = 9}$

4·8 Equations Involving Rational Expressions

1. A/an ________________ solution to a fractional equation is one that does not satisfy the equation because it results in division by __________.

2. To solve a fractional equation, multiply both sides of the equation by the _________ of the fractions in the equation.

Solve for the variable in each of the following equations.

3. (3) $\frac{3}{2x} + \frac{1}{3x} = \frac{11}{12}$

4. $\frac{2x}{3} - \frac{x}{2} = \frac{1}{4}$

5. $\frac{x}{x + 2} - \frac{4}{x + 1} = \frac{-2}{x + 2}$

6. $\frac{y + 1}{y + 7} = \frac{y - 4}{y - 1}$

7. $\frac{x - 3}{2x + 6} - \frac{x^2 + 6}{6x^2 - 54} = \frac{x + 3}{3x - 9}$

$$6(x+3)(x-3)\left[\frac{x-3}{2(x+3)} - \frac{(x^2+6)}{6(x+3)(x-3)}\right] = \left[\frac{x+3}{3(x+3)}\right]6(x+3)(x-3)$$

$$3(x-3)^2 - (x^2+6) = 2(x+3)^2$$

$$3x^2 - 18x + 27 - x^2 - 6 = 2x^2 + 12x + 18$$

$$-30x = -3$$

$$\boxed{x = \frac{1}{10}}$$

8. (47) $\frac{x + 2}{2x - 1} + \frac{x + 5}{x + 3} = \frac{5}{3}$

$$3(2x-1)(x+3)\left(\frac{x+2}{2x-1} + \frac{x+5}{x+3}\right) = \left(\frac{5}{3}\right)3(2x-1)(x+3)$$

$$3(x+3)(x+2) + 3(2x-1)(x+5) = 5(2x-1)(x+3)$$

$$3x^2 + 15x + 18 + 6x^2 + 27x - 15 = 10x^2 + 25x - 15$$

$$9x^2 + 42x + 3 = 10x^2 + 25x - 15$$

$$0 = x^2 - 17x - 18$$

$$0 = (x-18)(x+1)$$

$$\boxed{x = 18 \quad \text{OR} \quad x = -1}$$

9. $\frac{3x - 5}{5x - 5} + \frac{5x - 1}{7x - 7} - \frac{x - 4}{1 - x} = 2$

$$35(x-1)\left[\frac{3x-5}{5(x-1)} + \frac{5x-1}{7(x-1)} + \frac{x-4}{x-1}\right] = [2] \cdot 35(x-1)$$

$$7(3x-5) + 5(5x-1) + 35(x-4) = 70(x-1)$$

$$21x - 35 + 25x - 5 + 35x - 140 = 70x - 70$$

$$81x - 180 = 70x - 70$$

$$11x = 110$$

$$\boxed{x = 10}$$

10. (53) $\frac{x}{3x - 2} - \frac{8}{2x + 3} = \frac{2x^2 - 24x + 18}{6x^2 + 5x - 6}$

$$(3x-2)(2x+3)\left[\frac{x}{3x-2} - \frac{8}{2x+3}\right] = \left[\frac{2x^2 - 24x + 18}{(3x-2)(2x+3)}\right](3x-2)(2x+3)$$

$$x(2x+3) - 8(3x-2) = 2x^2 - 24x + 18$$

$$2x^2 + 3x - 24x + 16 = 2x^2 - 24x + 18$$

$$3x = 2$$

$x = \frac{2}{3}$ extraneous

$\boxed{\text{No solution}}$

7. $\frac{x - 3}{2x + 6} - \frac{x^2 + 6}{6x^2 - 54} = \frac{x + 3}{3x - 9}$

8. 47 $\frac{x + 2}{2x - 1} + \frac{x + 5}{x + 3} = \frac{5}{3}$

9. $\frac{3x - 5}{5x - 5} + \frac{5x - 1}{7x - 7} - \frac{x - 4}{1 - x} = 2$

10. 53 $\frac{x}{3x - 2} - \frac{8}{2x + 3} = \frac{2x^2 - 24x + 18}{6x^2 + 5x - 6}$

4·9 Applications

Find the solutions to the following problems.

1.
3

What number must be subtracted from both the numerator and the denominator of 3/5 to obtain 3/4? $X =$ the number

$$(4)(5-x)\left(\frac{3-x}{5-x}\right) = \left(\frac{3}{4}\right)(4)(5-x)$$

$$4(3-x) = 3(5-x)$$

$$12 - 4x = 15 - 3x$$

$$\boxed{-3 = x}$$

2.
13

Jolene can mow her yard in 2 hours. Her son Danny can mow the yard in 3 hours. Working together, how long will it take them to mow the yard? $x =$ # of hours to do the job together.

Jolene can do $\frac{1}{2}$ the job in 1 hr.
Danny can do $\frac{1}{3}$ the job in 1 hr.
Together, they can do $\frac{1}{x}$ the job in 1 hr.

$$6x\left(\frac{1}{2} + \frac{1}{3}\right) = \left(\frac{1}{x}\right)6x$$

$$3x + 2x = 6$$

$$5x = 6$$

$$x = \frac{6}{5} =$$

$$\boxed{1\tfrac{1}{5} \text{ hours together}}$$

4·9 Applications

Find the solutions to the following problems.

1. 3

What number must be subtracted from both the numerator and the denominator of 3/5 to obtain 3/4?

$$4(5-x)\ \frac{3-x}{5-x} = \frac{3}{4}\ 4(5-x) \qquad \frac{3-(-3)}{5-(-3)} = \frac{3}{4}$$

$$4(3-x) = 3(5-x) \qquad \frac{6}{8} = \frac{3}{4}$$

$$12 - 4x = 15 - 3x$$

$$-12 + 3x = -12 + 3x$$

$$-x = 3$$

$$x = -3$$

2. 13

Jolene can mow her yard in 2 hours. Her son Danny can mow the yard in 3 hours. Working together, how long will it take them to mow the yard?

		1 hr
Jolene	2	$\frac{1}{2}$
Danny	3	$\frac{1}{3}$
Together	x	$\frac{1}{x}$

$$(6x)\ \frac{1}{3} + \frac{1}{2}(6x) = \frac{1}{x}(6x)$$

$$2x + 3x = 6$$

$$\frac{5x}{5} = \frac{6}{5}$$

$$x = \frac{6}{5}$$

3. | 25

Tom's rowboat can travel 7 miles downstream in the same time (NOTE) that it can travel 3 miles upstream. If the speed of the current is 1 mph, what is Tom's rowing speed in still water?

x = Tom's speed in still water

$x+1$ = Tom's speed downstream (with the current)

$x-1$ = Tom's speed upstream (against the current)

Time upstream $= d \div r = 3 \div (x-1) = \frac{3}{x-1}$

Time downstream $= d \div r = 7 \div (x+1) = \frac{7}{x+1}$

Time upstream = Time downstream

$$(x+1)(x-1)\left(\frac{3}{x-1}\right) = \left(\frac{7}{x+1}\right)(x+1)(x-1)$$

$$3(x+1) = 7(x-1)$$

$$3x+3 = 7x-7$$

$$10 = 4x$$

$x =$ **$2\frac{1}{2}$ mph in still water**

4. | 9

The sum of the reciprocals of two consecutive integers is 7/12. What are the integers?

x = smaller integer

$x+1$ = larger integer

$\frac{1}{x}$ = reciprocal

$\frac{1}{x+1}$ = reciprocal

$$12x(x+1)\left(\frac{1}{x}+\frac{1}{x+1}\right) = \left(\frac{7}{12}\right)12x(x+1)$$

$$12(x+1)+12x = 7x(x+1)$$

$$12x+12+12x = 7x^2+7x$$

$$0 = 7x^2-17x-12$$

$$0 = (7x+4)(x-3)$$

$x = \frac{-4}{7}$ Discard because it is not an integer. OR $x = 3$

The integers are 3 and 4.

3. 25 Tom's rowboat can travel 7 miles downstream in the same time that it can travel 3 miles upstream. If the speed of the current is 1 mph, what is Tom's rowing speed in still water?

	d	= r	· t
Down	7	$x+1$	$\frac{7}{x+1}$
up	3	$x-1$	$\frac{3}{x-1}$
Still water		x	

$\frac{d}{r} = t$

time up = time down

$\frac{7}{x+1} = \frac{3}{x-1}$

$3(x+1) = 7(x-1)$

$3x+3 = 7x-7$

$10 = 4x$

$x = \frac{5}{2} = 2\frac{1}{2}$ mph in still water

4. 9 The sum of the reciprocals of two consecutive integers is 7/12.

What are the integers? $x,\ x+1$

$12x(x+1)\ \frac{1}{x} + \frac{1}{x+1} = \frac{7}{12}$

$12(x+1) + 12x = 7x(x+1)$

$12x + 12 + 12x = 7x^2 + 7x$

$24x + 12 = 7x^2 + 7x$

$12 = 7x^2 - 17x$

$0 = 7x^2 - 17x - 12$

$= (7x+4)(x-3)$

$x = -\frac{4}{7}$ or 3

3, 4

$\frac{1}{3} + \frac{1}{4} = \frac{7}{12}$

$4 + 3 = 7$

5. **23** One pipe can fill an oil tank 1 hour quicker than another pipe. The tanker trucks can empty the tank in 3 hours. With both pipes open and the tanker trucks draining the tank, it takes 2 hours to fill the tank. How long does it take each pipe to fill the tank?

x = time for pipe #1
$x+1$ = time for pipe #2
3 = time for tanker

$$6x(x+1)\left(\frac{1}{x}+\frac{1}{x+1}-\frac{1}{3}\right)=\left(\frac{1}{2}\right)6x(x+1)$$

$$6x+6+6x-2x^2-2x=3x^2+3x$$

$$0=5x^2-7x-6$$

$$0=(5x+3)(x-2)$$

$x=\frac{-3}{5}$ Since time cannot be negative, this answer is discarded. OR

$x = 2$ hrs. for pipe #1
$x+1 = 3$ hrs. for pipe #2

6. **31** Roachman climbs to the top of a cliff that is 3/4 mile high. He then parachutes down. His parachuting speed is 4 mph faster than his climbing speed. His total climbing and parachuting time is 1 2/3 hours. What is Roachman's climbing time?

x = climbing time
$\frac{5}{3}-x$ = parachuting time
climbing rate = $d \div t = \frac{3}{4x}$
parachuting rate = $d \div t = \frac{3}{4(\frac{5}{3}-x)}$

Parachuting rate = 4 + climbing rate

$$4x\left(\tfrac{5}{3}-x\right)\left[\frac{3}{4(\frac{5}{3}-x)}\right]=\left[4+\frac{3}{4x}\right]4x\left(\tfrac{5}{3}-x\right)$$

$$3x=16x\left(\tfrac{5}{3}-x\right)+3\left(\tfrac{5}{3}-x\right)$$

$$3x=\frac{80x}{3}-16x^2+5-3x$$

$$9x=80x-48x^2+15-9x$$

$$48x^2-62x-15=0$$

$$(2x-3)(24x+5)=0$$

$x=\frac{3}{2}$ hr. OR $x=\frac{-5}{24}$ Discard because time cannot be negative.

5. **23** One pipe can fill an oil tank 1 hour quicker than another pipe. The tanker trucks can empty the tank in 3 hours. With both pipes open and the tanker trucks draining the tank, it takes 2 hours to fill the tank. How long does it take each pipe to fill the tank?

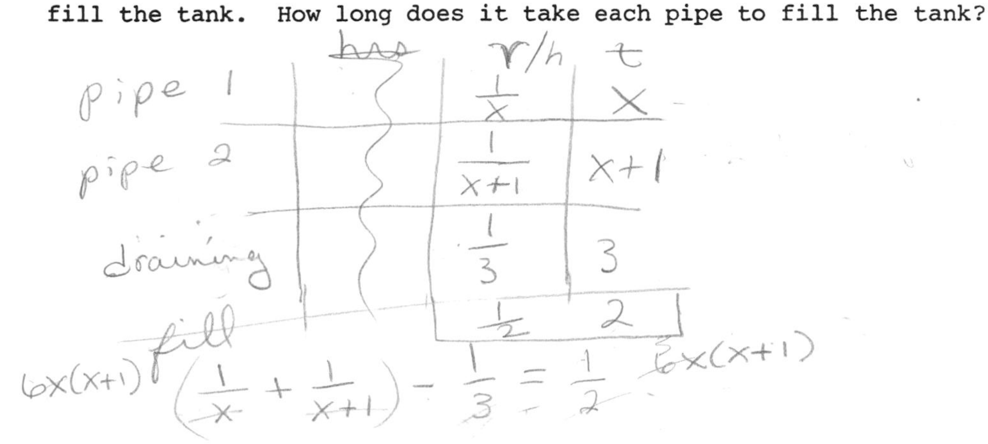

2 hrs #1
3 hrs #2

6. **31** Roachman climbs to the top of a cliff that is 3/4 mile high. He then parachutes down. His parachuting speed is 4 mph faster than his climbing speed. His total climbing and parachuting time is 1 2/3 hours. What is Roachman's climbing time?

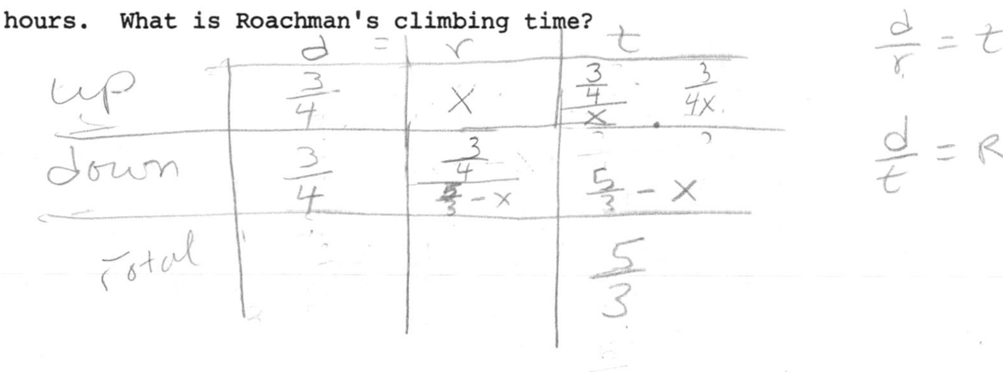

$\frac{5}{3} = x+4+x$

Chapter 4 Self-Test

Simplify the following rational expressions and reduce to lowest terms if possible.

1. $(6x^4y^7)(3x^2y)^{-2}$

2. $\dfrac{2x^2 + x - 10}{4 - x^2}$

3. $\left(\dfrac{3x^{-2}y}{x^0y^{-2}}\right)^{-3} \cdot \left(\dfrac{2x^3y^{-3}}{9x^{-4}y^2}\right)^{-2}$

4. $\dfrac{8x^3 + 1}{4x^2 - 1}$

5. $\dfrac{2a^2 - 13a + 15}{4a^2 - 9} \cdot \dfrac{2a - 1}{a - 5} \cdot \dfrac{2a - 1}{2a + 1}$

6. $\dfrac{4}{3x^2y} - \dfrac{2x}{xy^2} + \dfrac{y}{6xy}$

7. $\dfrac{4}{x^2 - 3x} + \dfrac{5}{9 - x^2}$

8. $\dfrac{\dfrac{3 - x}{3 + x}}{\dfrac{x - 3}{4x}}$

9. $\dfrac{3x^{-1} + 12x^{-2}}{1 + 4x^{-1}}$

10. $x - 5\overline{)x^2 - 8x + 20}$

Solve each equation.

11. $\dfrac{2}{5x} + \dfrac{1}{3x} = 2$

12. $\dfrac{x}{x + 3} - \dfrac{3}{x + 1} = \dfrac{-5}{x + 3}$

13. $\dfrac{6x}{3x + 2} + \dfrac{1}{x - 4} = \dfrac{-28x}{3x^2 - 10x - 8}$

14. $\dfrac{1}{x} + \dfrac{1}{x + 1} = \dfrac{7}{12}$

15. John can overhaul a lawn mower in 6 hours. Sam can do the same job in 2 hours. If they work together, how long will it take them to do the job?

C H A P T E R

5 Exponential and Radical Expressions

5 · 1 Rational Exponents

1. (a) It is known that $2^4 = 16$.

$$\text{So } (2^4)^{1/4} = (16)^{1/4}$$
$$2^{4 \cdot 1/4} = 16^{1/4}$$
$$2^1 = 16^{1/4}$$
$$2 = 16^{1/4}$$

Therefore, 2 is the principal 4th root of 16 since $16^{1/4} = 2$ and $2^4 = 16$.

(b) It is known that $(-2)^3 = -8$

$$\text{So } [(-2)^3]^{1/3} = [-8]^{1/3}$$
$$(-2)^{3 \cdot 1/3} = (-8)^{1/3}$$
$$(-2)^1 = (-8)^{1/3}$$
$$-2 = (-8)^{1/3}$$

Therefore, -2 is the third root of -8 since $(-8)^{1/3} = -2$ and $(-2)^3 = -8$.

CHAPTER

5 Exponential and Radical Expressions

5·1 Rational Exponents

1. (a) It is known that $2^4 = 16$.

$$\text{So } (2^4)^{1/4} = (16)^{1/4}$$
$$2^{4\cdot 1/4} = 16^{1/4}$$
$$2^1 = 16^{1/4}$$
$$2 = 16^{1/4}$$

Therefore, 2 is the principal 4th root of 16 since $16^{1/4} = 2$ and $2^4 = 16$.

(b) It is known that $(-2)^3 = -8$

$$\text{So } [(-2)^3]^{1/3} = [-8]^{1/3}$$
$$(-2)^{3\cdot 1/3} = (-8)^{1/3}$$
$$(-2)^1 = (-8)^{1/3}$$
$$-2 = (-8)^{1/3}$$

Therefore, -2 is the third root of -8 since $(-8)^{1/3} = -2$ and $(-2)^3 = -8$.

2. (a) Remember that $a^{m/n} = (a^{1/n})^m$ or $(a^m)^{1/n}$.

Therefore, $8^{2/3} = (8^{1/3})^2 = \underline{2}^{\,2} = 4$

or $8^{2/3} = (8^2)^{1/3} = (\underline{64})^{1/3} = \boxed{4}$

(b) Remember that $a^{-n} = 1/(a^n)$.

Therefore, $(27)^{-4/3} = \dfrac{1}{27^{4/3}}$

$= \dfrac{1}{(27^{1/3})^4}$

$= \dfrac{1}{3^4}$

$= \boxed{\dfrac{1}{81}}$

Evaluate each of the following:

3. $\left(\frac{36}{64}\right)^{1/2} = \dfrac{36^{\frac{1}{2}}}{64^{\frac{1}{2}}} = \dfrac{6}{8} = \boxed{\dfrac{3}{4}}$

4. $-1^{1/5} = \boxed{-1}$

5. | 11 — $(-1)^{1/2}$ This is not a real number because no real number squared equals -1.

6. $\left(\frac{1}{49}\right)^{3/2} = \left[\left(\frac{1}{49}\right)^{\frac{1}{2}}\right]^3 = \left(\frac{1}{7}\right)^3 = \boxed{\dfrac{1}{343}}$

7. $\left(-\frac{1}{64}\right)^{2/3} = \left[\left(\frac{-1}{64}\right)^{\frac{1}{3}}\right]^2 = \left[\frac{-1}{4}\right]^2 = \boxed{\dfrac{1}{16}}$

8. $(-32)^{-2/5} = \left[(-32)^{\frac{1}{5}}\right]^{-2} = (-2)^{-2} = \dfrac{1}{(-2)^2} = \boxed{\dfrac{1}{4}}$

9. $-\left(\frac{16}{81}\right)^{-3/4} = -\left[\left(\frac{16}{81}\right)^{\frac{1}{4}}\right]^{-3} = -\left[\frac{2}{3}\right]^{-3} = -\left(\frac{3}{2}\right)^3 = \boxed{-\dfrac{27}{8}}$

2. (a) Remember that $a^{m/n} = (a^{1/n})^m$ or $(a^m)^{1/n}$.

Therefore, $8^{2/3} = (8^{1/3})^2 = \underline{2}^{\,2} = 4$

or $8^{2/3} = (8^2)^{1/3} = (\underline{64})^{1/3} = 4$

(b) Remember that $a^{-n} = 1/(a^n)$.

Therefore, $(27)^{-4/3} = \dfrac{1}{27^{4/3}}$

$= \dfrac{1}{(27^{1/3})^4}$

$= \dfrac{1}{3^4}$

$=$

Evaluate each of the following:

3. $\left(\frac{36}{64}\right)^{1/2}$ $\frac{6}{8} = \frac{3}{4}$

4. $-1^{1/5}$ $\sqrt[5]{-1} = -1$

5. $(-1)^{1/2}$ not real

11

6. $\left(\frac{1}{49}\right)^{3/2}$ $\frac{1}{7}^3 = \frac{1}{343}$

7. $\left(-\frac{1}{64}\right)^{2/3}$ $-\frac{1}{4^2} = \frac{1}{16}$

8. $(-32)^{-2/5}$ $[(-32)^{\frac{1}{5}}]^{-2} = (-2)^{-2} = \frac{1}{-2^2} = \frac{1}{4}$

9. $-\left(\frac{16}{81}\right)^{-3/4}$ $= -\left(\frac{2}{3}\right)^{-3} = -\left(\frac{3^3}{2^3}\right) = \frac{27}{8}$

10. $\left(\frac{243}{1024}\right)^{2/5} = \left[\left(\frac{243}{1024}\right)^{\frac{1}{5}}\right]^2 = \left(\frac{3}{4}\right)^2 = \boxed{\frac{9}{16}}$

Simplify each of the following. Remember to write your answers using only positive exponents. Assume that all variables represent positive real numbers.

11. | 71 $\frac{9^{1/6} \cdot 9^{-2}}{9^{2/3}} = 9^{\frac{1}{6}+-2-\frac{2}{3}} = 9^{-\frac{5}{2}} = \left[\left(\frac{1}{9}\right)^{\frac{1}{2}}\right]^5 = \left(\frac{1}{3}\right)^5 = \boxed{\frac{1}{243}}$

12. $(4x^5)^{1/2}\,(27x^6)^{1/3} = 4^{\frac{1}{2}}x^{\frac{5}{2}} \cdot 27^{\frac{1}{3}}x^{\frac{6}{3}} = 2 \cdot 3 \cdot x^{\frac{5}{2}+2}$
$= \boxed{6x^{\frac{9}{2}}}$

13. | 91 $\frac{16^{3/8}x^{-1/2}y^{5/3}}{16^{1/8}x^{1/4}y^{1/2}} = 16^{\frac{3}{8}-\frac{1}{8}}x^{-\frac{1}{2}-\frac{1}{4}}y^{\frac{5}{3}-\frac{1}{2}} = 16^{\frac{1}{4}}x^{-\frac{3}{4}}y^{\frac{7}{6}} = \boxed{\frac{2y^{\frac{7}{6}}}{x^{\frac{3}{4}}}}$

14. $(x^{2/3} + 3)(x^{2/3} - 5) = \boxed{x^{\frac{4}{3}} - 2x^{\frac{2}{3}} - 15}$

15. $\frac{8^{7/9}}{8^{4/9}} = 8^{\frac{7}{9}-\frac{4}{9}} = 8^{\frac{3}{9}} = 8^{\frac{1}{3}} = \boxed{2}$

16. $\frac{8^{2/3}}{16^{3/4}} = \frac{(8^{\frac{1}{3}})^2}{(16^{\frac{1}{4}})^3} = \frac{2^2}{2^3} = \boxed{\frac{1}{2}}$

17. $(25^{3/2}x^6y^{-12})^{-5/3} = 25^{-\frac{15}{6}}x^{-\frac{30}{3}}y^{\frac{60}{3}} = 25^{-\frac{5}{2}}x^{-10}y^{20} = \boxed{\frac{y^{20}}{3125x^{10}}}$

18. | 105 $7x^{-2/3}(3x^{-1/6} + 4xy^{3/4}) = 21x^{-\frac{5}{6}} + 28x^{\frac{1}{3}}y^{\frac{3}{4}} = \boxed{\frac{21}{x^{\frac{5}{6}}} + 28x^{\frac{1}{3}}y^{\frac{3}{4}}}$

10. $\left(\frac{243}{1024}\right)^{2/5}$ $\left[\left(\frac{243}{1024}\right)^{1/5}\right]^2 = \left(\frac{3}{4}\right)^2 = \frac{9}{16}$

Simplify each of the following. Remember to write your answers using only positive exponents. Assume that all variables represent positive real numbers.

11. | 71

$\frac{9^{1/6} \cdot 9^{-2}}{9^{2/3}} = 9^{1/6 - \frac{2}{1} - \frac{2}{3}} = 9^{-\frac{5}{2}} = \frac{1}{9^{\frac{5}{2}}} = \frac{1}{3^5} = \frac{1}{243}$

12. $(4x^5)^{1/2}\ (27x^6)^{1/3} = (2x^{5/2})(3x^{\frac{6}{3}}) = 6x^{5/2 + \frac{6}{3}} = 6x^{9/2}$

13. | 91

$\frac{16^{3/8}x^{-1/2}y^{5/3}}{16^{1/8}x^{1/4}y^{1/2}} = 16^{3/8 - 1/8}x^{-\frac{1}{2}-\frac{1}{4}}y^{5/3 - 1/2} = \frac{16^{1/4}y^{7/6}}{x^{3/4}} = \frac{2y^{7/6}}{x^{3/4}}$

14. $(x^{2/3} + 3)(x^{2/3} - 5) = x^{4/3} - 5x^{2/3} + 3x^{2/3} - 15$

$x^{4/3} - 2x^{2/3} - 15$

15. $\frac{8^{7/9}}{8^{4/9}} = 8^{7/9 - 4/9} = 8^{\frac{3}{9}} = 8^{1/3} = 2$

16. $\frac{8^{2/3}}{16^{3/4}} = \frac{2^2}{2^3} = 2^{2-3} = \frac{1}{2}$

17. $(25^{3/2}x^6y^{-12})^{-5/3}$ $25^{3/2 \cdot -\frac{5}{3}}x^{6 \cdot -\frac{5}{3}}y^{-12 \cdot -\frac{5}{3}}$

$25^{-\frac{15}{6}}x^{-\frac{30}{3}}y^{\frac{60}{3}} = \frac{y^{20}}{25^{\frac{5}{2}}x^{10}} = \frac{y^{20}}{3125x^{10}}$

5^5

18. | 105

$7x^{-2/3}(3x^{-1/6} + 4xy^{3/4})$

$7 \cdot 3x^{-2/3 + 1/6} + 7 \cdot 4x^{-2/3 + 1}y^{3/4}$

$21x^{-\frac{4}{6}+\frac{1}{6}} + 28x^{-\frac{2}{3}+\frac{3}{3}}y^{3/4}$

$21x^{-\frac{5}{6}}\ 28x^{1/3}y^{3/4} = \frac{21}{x^{\frac{5}{6}}} + 28x^{\frac{1}{3}}y^{\frac{3}{4}}$

5·2 Radicals

1. The symbol $\sqrt{\ }$ is called a radical sign.

2. In the expression $27^{1/3} = \sqrt[3]{27}$;

   (a) $\sqrt[3]{27}$ is read the third (or cube) root of 27.

   (b) The 27 is called the radicand.

   (c) The ^{3} is called the index or order of the radical.

   (d) If no index is stated, an expression such as $9^{1/2} = \sqrt{9}$ is read the square root of 9 and the index is understood to be 2.

Write the correct choice in the blank provided.

Answer	Problem	Choice
d	3. $\sqrt{a}\,\sqrt[4]{a}$	(a) $a^{7/3}$
i	4. $\dfrac{\sqrt{a^6}}{\sqrt{a^3}}$	(b) $\sqrt[4]{a}$
h	5. $\sqrt[4]{a^4}$	(c) $a^{1/6}$
g	6. $\sqrt[5]{a^5}$	(d) $a^{3/4}$
b	7. $a^{1/4}$	(e) $a^{3/7}$
f	8. $a^{2/5}$	(f) $\sqrt[5]{a^2}$
a	9. $(\sqrt[3]{a})^7$	(g) a
e	10. $\sqrt[7]{a^3}$	(h) $\lvert a \rvert$
c	11. $\sqrt[3]{\sqrt[4]{a^2}}$	(i) $\sqrt{a^3}$

Convert the following expressions into radical expressions.

12. $(-3x)^{7/3} = \boxed{\sqrt[3]{(-3x)^7} \text{ OR } (\sqrt[3]{-3x})^7}$

13. $\left(\frac{3}{8}\right)^{2/3} = \boxed{\sqrt[3]{\left(\frac{3}{8}\right)^2} \text{ OR } \left(\sqrt[3]{\frac{3}{8}}\right)^2}$

5·2 Radicals

1. The symbol $\sqrt{\ }$ is called a ______ ______.

2. In the expression $27^{1/3} = \sqrt[3]{27}$;

   (a) $\sqrt[3]{27}$ is read ______________.

   (b) The 27 is called the ______.

   (c) The ^{3} is called the ______ or ______ of the radical.

   (d) If no index is stated, an expression such as $9^{1/2} = \sqrt{9}$ is read ______________ and the index is understood to be ______.

Write the correct choice in the blank provided.

____	3. $\sqrt{a}\,\sqrt[4]{a}$	(a)	$a^{7/3}$
____	4. $\dfrac{\sqrt{a^6}}{\sqrt{a^3}}$	(b)	$\sqrt[4]{a}$
____	5. $\sqrt[4]{a^4}$	(c)	$a^{1/6}$
____	6. $\sqrt[5]{a^5}$	(d)	$a^{3/4}$
____	7. $a^{1/4}$	(e)	$a^{3/7}$
____	8. $a^{2/5}$	(f)	$\sqrt[5]{a^2}$
____	9. $\left(\sqrt[3]{a}\right)^7$	(g)	a
____	10. $\sqrt[7]{a^3}$	(h)	$\lvert a\rvert$
____	11. $\sqrt[3]{\sqrt[4]{a^2}}$	(i)	$\sqrt{a^3}$

Convert the following expressions into radical expressions.

12. $(-3x)^{7/3}$

13. $\left(\frac{3}{8}\right)^{2/3}$

14. 51 $(16x^2 - 25y^2)^{1/2} = \boxed{\sqrt{16x^2 - 25y^2}}$

15. $4^{-1/2} = \frac{1}{4^{\frac{1}{2}}} = \frac{1}{\sqrt{4}} = \boxed{\frac{1}{2}}$

Convert the following radical expressions into expressions with rational exponents and simplify. Assume that all variables represent positive real numbers.

16. 21 $\sqrt[3]{(-4)^3} = \left[(-4)^3\right]^{\frac{1}{3}} = \boxed{-4}$

17. $\sqrt{(-5)^4} = \left[(-5)^4\right]^{\frac{1}{4}} = (625)^{\frac{1}{4}} = \boxed{5}$ Since -5 is being raised to an even power, the value under the radical is positive.

18. $\sqrt{81} \cdot \sqrt[3]{8} = 9 \cdot 2 = \boxed{18}$

19. $\sqrt[5]{\sqrt{x^{15}}} = \left[(x^{15})^{\frac{1}{2}}\right]^{\frac{1}{5}} = \boxed{x^{\frac{3}{2}}}$

20. $\sqrt[5]{\frac{-1}{32}} = \left(\frac{-1}{32}\right)^{\frac{1}{5}} = \boxed{\frac{1}{-2}}$

21. $\frac{\sqrt[3]{125}}{\sqrt{9}} = \boxed{\frac{5}{3}}$

22. $\sqrt{(2x - 3y)^2} = \boxed{2x - 3y}$

23. 75 $\left(\sqrt[3]{(-2xy^3)}\right)^6 = \left[(-2xy^3)^{\frac{1}{3}}\right]^6 = (-2xy^3)^2 = \boxed{4x^2y^6}$

24. $\left(\sqrt{(2x + 1)}\right)^6 = \left[(2x+1)^{\frac{1}{3}}\right]^6 = (2x+1)^2 = \boxed{4x^2+4x+1}$

25. 81 $\sqrt{x^3}\sqrt[6]{x} = x^{\frac{3}{2}} \cdot x^{\frac{1}{6}} = x^{\frac{3}{2}+\frac{1}{6}} = x^{\frac{10}{6}} = \boxed{x^{\frac{5}{3}}}$

14. (51) $(16x^2 - 25y^2)^{1/2}$

15. $4^{-1/2}$

Convert the following radical expressions into expressions with rational exponents and simplify. Assume that all variables represent positive real numbers.

16. (21) $\sqrt[3]{(-4)^3}$

17. $\sqrt{(-5)^4}$

18. $\sqrt{81} \cdot \sqrt[3]{8}$

19. $\sqrt[5]{\sqrt{x^{15}}}$

20. $\sqrt[5]{\dfrac{-1}{32}}$

21. $\dfrac{\sqrt[3]{125}}{\sqrt{9}}$

22. $\sqrt{(2x - 3y)^2}$

23. (75) $\left(\sqrt[3]{(-2xy^3)}\right)^6$

24. $\left(\sqrt[3]{(2x + 1)}\right)^6$

25. (81) $\sqrt{x^3}\,\sqrt[6]{x}$

5 · 3 Simplifying Radical Expressions

1. (a) The expression equivalent to $\sqrt[n]{ab}$ is written $\sqrt[n]{a} \cdot \sqrt[n]{b}$.

   (b) The expression equivalent to $\sqrt[n]{\frac{a}{b}}$ is written $\frac{\sqrt[n]{a}}{\sqrt[n]{b}}$.

Simplify the following radical expressions. Assume that all variables represent positive real numbers.

2. $\sqrt[3]{-4} \cdot \sqrt[3]{16} = \sqrt[3]{-64} = \boxed{-4}$

3. $\sqrt{72x^3y^5} = \sqrt{36x^2y^4} \cdot \sqrt{2xy} = \boxed{6xy^2\sqrt{2xy}}$

4. $\sqrt[3]{81x^2y^4z^6} = \sqrt[3]{81y^3z^6} \cdot \sqrt[3]{x^2y} = \boxed{3yz^2\sqrt[3]{x^2y}}$

5. $\sqrt[4]{32x^5y^7} = \sqrt[4]{16x^4y^4} \cdot \sqrt[4]{2xy^3} = \boxed{2xy\sqrt[4]{2xy^3}}$

   47

6. $\sqrt[3]{\frac{27x^4}{8y^6}} = \frac{\sqrt[3]{27x^4}}{\sqrt[3]{8y^6}} = \frac{\sqrt[3]{27x^3} \cdot \sqrt[3]{x}}{2y^2} = \boxed{\frac{3x\sqrt[3]{x}}{2y^2}}$

7. $\frac{\sqrt{7x^5y}}{\sqrt{28xy^6}} = \frac{\sqrt{x^4} \cdot \sqrt{7xy}}{\sqrt{4y^6} \cdot \sqrt{7x}} = \frac{x^2\sqrt{7xy}}{2y^3\sqrt{7x}} = \frac{x^2}{2y^3} \cdot \sqrt{\frac{7xy}{7x}} = \boxed{\frac{x^2\sqrt{y}}{2y^3}}$

8. $\frac{\sqrt{-4x^2}}{\sqrt{-16x^4}} =$ Can't be done; the square root of a negative number is not a real number.

5·3 Simplifying Radical Expressions

1. (a) The expression equivalent to $\sqrt[n]{ab}$ is written $\sqrt[n]{a} \cdot \sqrt[n]{b}$.

   (b) The expression equivalent to $\sqrt[n]{\dfrac{a}{b}}$ is written $\dfrac{\sqrt[n]{a}}{\sqrt[n]{b}}$.

Simplify the following radical expressions. Assume that all variables represent positive real numbers.

2. $\sqrt[3]{-4} \cdot \sqrt[3]{16}$ $\sqrt[3]{-64}$ $-4^{1/3} \cdot 16^{1/3} = -64^{1/3}$ $\sqrt[3]{-64}$ $= -4$

3. $\sqrt{72x^3y^5}$ $\sqrt{36x^2y^4}\sqrt{2xy} = 6xy^2\sqrt{2xy}$

4. $\sqrt[3]{81x^2y^4z^6}$ $3yz^2\sqrt[3]{3x^2y}$ $= \sqrt{27}\sqrt{3}$

5. $\sqrt[4]{32x^5y^7}$ $=$ ~~$\sqrt[4]{16 \cdot 2 \cdot x \cdot x \cdot x \cdot x \cdot x \cdot y \cdot y \cdot y \cdot y \cdot y \cdot y \cdot y}$~~ $\sqrt[4]{2xy^3}$

   $2xy\sqrt[4]{2xy^3}$

6. $\sqrt[3]{\dfrac{27x^4}{8y^6}}$ $= \dfrac{\sqrt[3]{27x^4}}{\sqrt[3]{8y^6}} = \dfrac{3x\sqrt[3]{x}}{2y^2}$

7. $\dfrac{\sqrt{7x^5y}}{\sqrt{28xy^6}}$ $= \dfrac{x^2\sqrt{7xy}}{2y^3\sqrt{7x}} = \dfrac{\sqrt{7x}}{\sqrt{7x}} = \dfrac{x^2\sqrt{7xy}\,(\sqrt{7x})}{2y^3\sqrt{7x}\,(\sqrt{7x})}$

   $x^2\sqrt{49x^2y} = \dfrac{7x^3\sqrt{y}}{14xy^3}$

   $2y^3\sqrt{49x^2}$

   $2y^3 \cdot 7x$

   $14xy^3$

   $\dfrac{x^2\sqrt{y}}{2y^3}$

8. $\dfrac{\sqrt{-4x^2}}{\sqrt{-16x^4}}$ $= \dfrac{\sqrt{-4x^2}}{\sqrt{-16x^4}}$ not a real #

9. $\sqrt[3]{\frac{-125}{8y^9}} = \frac{\sqrt[3]{-125}}{\sqrt[3]{8y^9}} = \boxed{\frac{-5}{2y^3}}$

10. $\frac{\sqrt[3]{(x+y)^6 z^{10}}}{\sqrt[3]{(x+y)^4 z}} = \sqrt[3]{\frac{(x+y)^6 z^{10}}{(x+y)^4 z}} = \sqrt[3]{(x+y)^2 z^9} = \boxed{z^3\sqrt[3]{(x+y)^2}}$

11. | 81 $\sqrt[3]{\frac{7}{8x}} \cdot \sqrt[3]{\frac{x^2}{x^2}} = \sqrt[3]{\frac{7x^2}{8x^3}} = \frac{\sqrt[3]{7x^2}}{\sqrt[3]{8x^3}} = \frac{\sqrt[3]{7x^2}}{2x}$

12. | 73 $\sqrt{\frac{2x}{3y^3}} \cdot \sqrt{\frac{3y}{3y}} = \sqrt{\frac{6xy}{9y^4}} = \boxed{\frac{\sqrt{6xy}}{3y^2}}$

---

5 · 4 Operations with Radical Expressions

1. "Like radicals" are radicals that have the same index or order and radicand.

2. The conjugate of $(2 - \sqrt{3})$ is $2+\sqrt{3}$. The conjugate is used to rationalize the denominator of radical expressions.

Perform the indicated operations. Assume that all variables represent positive real numbers.

3. | 9 $5\sqrt{12} - 2\sqrt{27} + 3\sqrt{3} =$

$5\sqrt{4}\cdot\sqrt{3} - 2\sqrt{9}\cdot\sqrt{3} + 3\sqrt{3} =$

$10\sqrt{3} - 6\sqrt{3} + 3\sqrt{3} =$

$\boxed{7\sqrt{3}}$

4. $\sqrt[3]{54x^4} + 3x\sqrt[3]{16x} =$

$\sqrt[3]{27x^3}\cdot\sqrt[3]{2x} + 3x\sqrt[3]{8}\cdot\sqrt[3]{2x} =$

$3x\sqrt[3]{2x} + 3x\cdot 2\sqrt[3]{2x} =$

$\boxed{9x\sqrt[3]{2x}}$

9. $\sqrt[3]{\dfrac{-125}{8y^9}}$ $= \dfrac{\sqrt[3]{-125}}{\sqrt[3]{8y^9}} = \dfrac{-5}{2y^3}$

10. $\dfrac{\sqrt[3]{(x + y)^6 z^{10}}}{\sqrt[3]{(x + y)^4 z}}$ $\sqrt[3]{\dfrac{(x+y)^6 z^{10}}{(x+y)^4 z}}$ $\sqrt[3]{(x+y)^2 z^9}$ $z^3\sqrt[3]{(x+y)^2}$

11. [81] $\sqrt[3]{\dfrac{7}{8x}}$ $\dfrac{\sqrt[3]{7}}{\sqrt[3]{8x}} \cdot \dfrac{\sqrt[3]{x^2}}{\sqrt[3]{x^2}} = \dfrac{\sqrt[3]{7x^2}}{\sqrt[3]{8x^3}} = \dfrac{\sqrt[3]{7x^2}}{2x}$

12. [73] $\sqrt{\dfrac{2x}{3y^3}}$ $\dfrac{\sqrt{2x}}{\sqrt{3y^3}} \cdot \dfrac{\sqrt{3y}}{\sqrt{3y}} = \dfrac{\sqrt{6xy}}{\sqrt{9y^4}} = \dfrac{\sqrt{6xy}}{3y^2}$

5·4 Operations with Radical Expressions

1. "Like radicals" are radicals that have the same index and radicand.

2. The conjugate of $(2 - \sqrt{3})$ is $(2+\sqrt{3})$. The conjugate is used to rationalize the denominator of radical expressions.

Perform the indicated operations. Assume that all variables represent positive real numbers.

3. [9] $5\sqrt{12} - 2\sqrt{27} + 3\sqrt{3}$

$5\sqrt{4}\sqrt{3} - 2\sqrt{9}\sqrt{3} + 3\sqrt{3}$

$10\sqrt{3} - 6\sqrt{3} + 3\sqrt{3}$

$7\sqrt{3}$

4. $\sqrt[3]{54x^4} + 3x\sqrt[3]{16x}$

$\sqrt[3]{27x^3}\sqrt[3]{2x} + 3x\sqrt[3]{8}\sqrt[3]{2x}$

$3x\sqrt[3]{2x} + 6x\sqrt[3]{2x}$

$9x\sqrt[3]{2x}$

5\. (25) $\sqrt{27x} + 2\sqrt{12x} - \sqrt{150y} - 4\sqrt{24y} =$

$\sqrt{9}\cdot\sqrt{3x} + 2\sqrt{4}\cdot\sqrt{3x} - \sqrt{25}\cdot\sqrt{6y} - 4\sqrt{4}\cdot\sqrt{6y} =$

$3\sqrt{3x} + 4\sqrt{3x} - 5\sqrt{6y} - 8\sqrt{6y} =$

$\boxed{7\sqrt{3x} - 13\sqrt{6y}}$

6\. $\frac{3}{\sqrt{2}} + 8\sqrt{2} = \frac{3}{\sqrt{2}}\cdot\frac{\sqrt{2}}{\sqrt{2}} + 8\sqrt{2} =$

$\frac{3\sqrt{2}}{2} + 8\sqrt{2} = \frac{3}{2}\sqrt{2} + 8\sqrt{2} =$

$\boxed{\frac{19}{2}\sqrt{2}}$

7\. (51) $\frac{1}{\sqrt[3]{2}} - \sqrt[3]{108} =$

$\frac{1}{\sqrt[3]{2}}\cdot\frac{\sqrt[3]{4}}{\sqrt[3]{4}} - \sqrt[3]{27}\cdot\sqrt[3]{4} =$

$\frac{\sqrt[3]{4}}{\sqrt[3]{8}} - 3\sqrt[3]{4} = \frac{1}{2}\sqrt[3]{4} - 3\sqrt[3]{4} =$

$\boxed{-\frac{5}{2}\sqrt[3]{4}}$

8\. $\sqrt{\frac{2}{5}} + 3\sqrt{10} - \sqrt{\frac{5}{2}} =$

$\frac{\sqrt{2}}{\sqrt{5}}\cdot\frac{\sqrt{5}}{\sqrt{5}} + 3\sqrt{10} - \frac{\sqrt{5}}{\sqrt{2}}\cdot\frac{\sqrt{2}}{\sqrt{2}} =$

$\frac{\sqrt{10}}{5} + 3\sqrt{10} - \frac{\sqrt{10}}{2} =$

$\frac{2\sqrt{10}}{10} + \frac{30\sqrt{10}}{10} - \frac{5\sqrt{10}}{10} = \boxed{\frac{27}{10}\sqrt{10}}$

5.5 More Operations with Radical Expressions

1\. $(3\sqrt{6} + \sqrt{2})(3\sqrt{6} - \sqrt{2}) =$

$9\sqrt{36} - \sqrt{4} =$

$9\cdot 6 - 2 =$

$\boxed{52}$

2\. $(2\sqrt{5} - 1)^2 =$

$4\sqrt{25} - 4\sqrt{5} + 1 =$

$20 - 4\sqrt{5} + 1 =$

$\boxed{21 - 4\sqrt{5}}$

5. [25] $\sqrt{27x} + 2\sqrt{12x} - \sqrt{150y} - 4\sqrt{24y}$

$\sqrt{9}\sqrt{3x} + 2\sqrt{4}\sqrt{3x} - \sqrt{25}\sqrt{6y} - 4\sqrt{4}\sqrt{6y}$

$3\sqrt{3x} + 4\sqrt{3x} - 5\sqrt{6y} - 8\sqrt{6y}$

$7\sqrt{3x} - 13\sqrt{6y}$

6. $\dfrac{3}{\sqrt{2}} + 8\sqrt{2}$

$\dfrac{3}{\sqrt{2}} \cdot \dfrac{\sqrt{2}}{\sqrt{2}} + 8\sqrt{2} = \dfrac{3\sqrt{2}}{\sqrt{4}} + 8\sqrt{2}$

$\dfrac{3\sqrt{2}}{2} + \dfrac{8\sqrt{2}}{1} \cdot \dfrac{2}{2}$

$\dfrac{3\sqrt{2}}{2} + \dfrac{16\sqrt{2}}{2} = \dfrac{19\sqrt{2}}{2}$

$\dfrac{19}{2}\sqrt{2}$

7. [51] $\dfrac{1}{\sqrt[3]{2}} - \sqrt[3]{108}$

$\dfrac{1}{\sqrt[3]{2}} \cdot \dfrac{\sqrt[3]{4}}{\sqrt[3]{4}} - \dfrac{\sqrt[3]{108}}{1}$

$\dfrac{\sqrt[3]{4}}{\sqrt[3]{8}} - \dfrac{\sqrt[3]{27}\sqrt[3]{4}}{1}$

$\dfrac{\sqrt[3]{4}}{2} - \dfrac{3\sqrt[3]{4}}{1} \cdot \dfrac{2}{2} = \dfrac{\sqrt[3]{4} - 6\sqrt[3]{4}}{2} = -\dfrac{5}{2}\sqrt[3]{4}$

8. $\sqrt{\dfrac{2}{5}} + 3\sqrt{10} - \sqrt{\dfrac{5}{2}}$

5.5 More Operations with Radical Expressions

1. $(3\sqrt{6} + \sqrt{2})(3\sqrt{6} - \sqrt{2})$

$9\sqrt{36} - 3\sqrt{12} + 3\sqrt{12} - \sqrt{4}$

$9 \cdot 6 - 2$

$54 - 2 = 52$

2. $(2\sqrt{5} - 1)^2$ $\quad (2\sqrt{5} - 1)(2\sqrt{5} - 1)$

$4\sqrt{25} - 2\sqrt{5} - 2\sqrt{5} + 1$

$20 - 4\sqrt{5} + 1$

$21 - 4\sqrt{5}$

3. $(\sqrt[3]{2x^2} - 3x)(\sqrt[3]{4x} + 2) =$

$\sqrt[3]{8x^3} + 2\sqrt[3]{2x^2} - 3x\sqrt[3]{4x} - 6x =$

$2x + 2\sqrt[3]{2x^2} - 3x\sqrt[3]{4x} - 6x =$

$\boxed{-4x + 2\sqrt[3]{2x^2} - 3x\sqrt[3]{4x}}$

4. [49] $(\sqrt[3]{3} - 1)(\sqrt[3]{9} + \sqrt[3]{3} + 1) =$

$\sqrt[3]{27} + \sqrt[3]{9} + \sqrt[3]{3} - \sqrt[3]{9} - \sqrt[3]{3} - 1 =$

$\sqrt[3]{27} - 1 = 3 - 1 =$

$\boxed{2}$

5. [17] $(2\sqrt{x} + 3\sqrt{y})(4\sqrt{x} + 5\sqrt{y}) =$

$8\sqrt{x^2} + 10\sqrt{xy} + 12\sqrt{xy} + 15\sqrt{y^2} =$

$\boxed{8x + 22\sqrt{xy} + 15y}$

6. [59] $\dfrac{2 + \sqrt{3}}{2 - \sqrt{3}} \cdot \dfrac{2 + \sqrt{3}}{2 + \sqrt{3}} =$

$\dfrac{4 + 2\sqrt{3} + 2\sqrt{3} + \sqrt{9}}{4 - \sqrt{9}} =$

$\dfrac{4 + 4\sqrt{3} + 3}{4 - 3} = \boxed{7 + 4\sqrt{3}}$

7. $\dfrac{\sqrt{6} - \sqrt{3}}{\sqrt{6} + \sqrt{3}} \cdot \dfrac{\sqrt{6} - \sqrt{3}}{\sqrt{6} - \sqrt{3}} =$

$\dfrac{\sqrt{36} - 2\sqrt{18} - \sqrt{9}}{\sqrt{36} - \sqrt{9}} =$

$\dfrac{6 - 2\sqrt{9}\cdot\sqrt{2} - 3}{6 - 3} =$

$\dfrac{3 - 6\sqrt{2}}{3} = \dfrac{3(1 - 2\sqrt{2})}{3} =$

$\boxed{1 - 2\sqrt{2}}$

8. $\dfrac{x - \sqrt{3}}{2\sqrt{x} - 5} \cdot \dfrac{2\sqrt{x} + 5}{2\sqrt{x} + 5} =$

$\dfrac{2x\sqrt{x} + 5x - 2\sqrt{3x} - 5\sqrt{3}}{4\sqrt{x^2} - 25} =$

$\boxed{\dfrac{2x\sqrt{x} + 5x - 2\sqrt{3x} - 5\sqrt{3}}{4x - 25}}$

5.6 Radical Equations

1. Equations in which the variable appears in at least one radicand are called radical equations.

3. $(\sqrt[3]{2x^2} - 3x)(\sqrt[3]{4x} + 2)$

4. | 49 $(\sqrt[3]{3} - 1)(\sqrt[3]{9} + \sqrt[3]{3} + 1)$

5. | 17 $(2\sqrt{x} + 3\sqrt{y})(4\sqrt{x} + 5\sqrt{y})$

6. | 59 $\dfrac{2 + \sqrt{3}}{2 - \sqrt{3}}$

7. $\dfrac{\sqrt{6} - \sqrt{3}}{\sqrt{6} + \sqrt{3}}$

8. $\dfrac{x - \sqrt{3}}{2\sqrt{x} - 5}$

5.6 Radical Equations

1. Equations in which the variable appears in at least one radicand are called radical equations

2. Solve: $\sqrt{x+7} - 1 = x$

$\sqrt{x+7} = x + 1$ (1) Isolate the radical term.

$(\sqrt{x+7})^2 = (x+1)^2$ (2) Square both sides.

$x + 7 = x^2 + 2x + 1$ (3) Simplify.

$0 = x^2 + x - 6$ (4) Place in standard form.

$0 = (x+3)(x-2)$ (5) Factor and solve.

$x = -3$ or $x = 2$ (6) Check for extraneous solutions, which give a false statement of equality.

$\sqrt{-3+7} - 1 = -3$ $\sqrt{2+7} - 1 = 2$

$\sqrt{4} - 1 = -3$ $\sqrt{9} - 1 = 2$

$2 - 1 = -3$ $3 - 1 = 2$

$1 \neq -3$ $2 = 2$

Therefore, the only solution is 2.

Solve the following radical equations. Remember to check for extraneous solutions.

3. 17 $\sqrt{3x^2 - x - 6} = \sqrt{2x^2 + x + 9}$

$(\sqrt{3x^2 - x - 6})^2 = (\sqrt{2x^2 + x + 9})^2$

$3x^2 - x - 6 = 2x^2 + x + 9$

$x^2 - 2x - 15 = 0$

$(x-5)(x+3) = 0$

$x = {}^{+}5$ OR $x = -3$

No extraneous roots

4. $\sqrt{x^2 + 9} - 5 = 0$

$\sqrt{x^2+9} = 5$

$(\sqrt{x^2+9})^2 = 5^2$

$x^2 + 9 = 25$

$x^2 - 16 = 0$

$(x+4)(x-4) = 0$

$x = 4$ OR $x = -4$

No extraneous roots

2. Solve: $\sqrt{x + 7} - 1 = x$

$\sqrt{x + 7} = x + 1$	(1)	Isolate the radical term.
$(\sqrt{x + 7})^2 = (x + 1)^2$	(2)	Square both sides.
$x + 7 = x^2 + 2x + 1$	(3)	Simplify.
$0 = x^2 + x - 6$	(4)	Place in standard form.
$0 = (x - 2)(x + 3)$	(5)	Factor and solve.
$x = 2$ or $x = -3$	(6)	Check for extraneous solutions, which give a false statement of equality.

$\sqrt{2 + 7} - 1 = 2$ $\qquad$ $\sqrt{-3 + 7} - 1 = -3$

$\sqrt{9} - 1 = 2$ $\qquad$ $\sqrt{4} - 1 = -3$

$3 - 1 = 2$ $\qquad$ $2 - 1 = 1$

$1 \neq -3$ $\qquad$ $2 = 2$

Therefore, the only solution is 2.

Solve the following radical equations. Remember to check for extraneous solutions.

3. 17 $\sqrt{3x^2 - x - 6} = \sqrt{2x^2 + x + 9}$ $\qquad$ 4. $\sqrt{x^2 + 9} - 5 = 0$

5. $x = \sqrt{2x + 15}$

$x^2 = \left(\sqrt{2x+15}\right)^2$

$x^2 = 2x + 15$

$x^2 - 2x - 15 = 0$

$(x-5)(x+3) = 0$

$\boxed{x = 5}$ OR $x = -3$

Extraneous, because $-3 \neq \sqrt{-6+15}$

6. [33] $\sqrt{x + 8} - x = -4$

$\left(\sqrt{x+8}\right)^2 = (x-4)^2$

$x + 8 = x^2 - 8x + 16$

$0 = x^2 - 9x + 8$

$0 = (x-1)(x-8)$

$x = 1$ OR $\boxed{x = 8}$

Extraneous, because $\sqrt{1+8} - 1 \neq -4$

7. [47] $\sqrt{2x + 3} + \sqrt{x + 2} = 2$

$\left(\sqrt{2x+3}\right)^2 = \left(2 - \sqrt{x+2}\right)^2$

$2x + 3 = 4 - 4\sqrt{x+2} + \left(\sqrt{x+2}\right)^2$

$2x + 3 = 4 - 4\sqrt{x+2} + x + 2$

$(x-3)^2 = \left(-4\sqrt{x+2}\right)^2$

$x^2 - 6x + 9 = 16(x+2)$

$x^2 - 6x + 9 = 16x + 32$

$x^2 - 22x - 23 = 0$

$(x-23)(x+1) = 0$

$x = 23$ (Extraneous) OR $\boxed{x = -1}$

8. $2\sqrt[3]{x} - 6 = 0$

$2\sqrt[3]{x} = 6$

$\left(\sqrt[3]{x}\right)^3 = (3)^3$

$\boxed{x = 27}$

9. [51] $\left(\sqrt{x + 2} + \sqrt{x - 1}\right)^2 = \left(\sqrt{4x + 1}\right)^2$

$x + 2 + 2\sqrt{x+2}\cdot\sqrt{x-1} + x - 1 = 4x + 1$

$2x + 1 + 2\sqrt{x^2+x-2} = 4x + 1$

$\left(2\sqrt{x^2+x-2}\right)^2 = (2x)^2$

$4(x^2+x-2) = 4x^2$

$4x^2 + 4x - 8 = 4x^2$

$4x - 8 = 0$ $\boxed{x = 2}$

10. $\left(\sqrt[4]{2x - 1}\right)^4 = (2)^4$

$2x - 1 = 16$

$2x = 17$

$\boxed{x = \frac{17}{2}}$

5. $x = \sqrt{2x + 15}$

6. 33 $\sqrt{x + 8} - x = -4$

7. 47 $\sqrt{2x + 3} + \sqrt{x + 2} = 2$

8. $2\sqrt[3]{x} - 6 = 0$

9. 51 $\sqrt{x + 2} + \sqrt{x - 1} = \sqrt{4x + 1}$

10. $\sqrt[4]{2x - 1} = 2$

5.7 Complex Numbers

1. Any number that can be expressed in the form $a + bi$, where a and b are real numbers and $i = \sqrt{-1}$, is called a complex number. In the expression $a + bi$, a is called the real part and b is called the imaginary part of the complex number $a + bi$.

2. Write 1, -1, i, or $-i$ in the blank provided.

$i = i$ $i^{14} = (i^4)^3 \cdot i^2 = -1$

$i^3 = i^2 \cdot i = -i$ $i^{66} = (i^4)^{16} \cdot i^2 = -1$

$i^8 = (i^4)^2 = (+1)^2 = 1$ $i^{20} = (i^4)^5 = 1$

$i^{100} = (i^4)^{25} = 1$ $i^{33} = (i^4)^8 \cdot i = i$

$i^{37} = (i^4)^9 \cdot i = i$ $i^{43} = (i^4)^{10} \cdot i^3 = -i$

Remember:
$i = i$
$i^2 = -1$
$i^3 = -i$
$i^4 = 1$

3. $5 - 2i$ and $5 + 2i$ are called complex conjugates of each other.

Simplify and express your answer in the form $a + bi$.

4. $(\sqrt{-36})(\sqrt{-100}) =$

$(\sqrt{36} \cdot \sqrt{-1})(\sqrt{100} \cdot \sqrt{-1}) =$
$6i \cdot 10i = 60i^2 = -60 =$
$\boxed{-60 + 0i}$

5. (17) $3\sqrt{18} - 2\sqrt{-24} =$

$3\sqrt{9} \cdot \sqrt{2} - 2\sqrt{4} \cdot \sqrt{6} \cdot \sqrt{-1} =$
$\boxed{9\sqrt{2} - 4\sqrt{6}\, i}$

6. $\dfrac{(\sqrt{-6})(\sqrt{-12})}{\sqrt{-16}} = \dfrac{i\sqrt{6} \cdot i\sqrt{12}}{4i} =$

$\dfrac{i^2\sqrt{72}}{4i} = \dfrac{-\sqrt{36} \cdot \sqrt{2}}{4i} = \dfrac{-6\sqrt{2}}{4i} =$

$\dfrac{-3\sqrt{2}}{2i} \cdot \dfrac{i}{i} = \dfrac{3i\sqrt{2}}{-2} = \boxed{0 + \dfrac{-3\sqrt{2}}{2}\, i}$

7. $-\sqrt{50} + 2\sqrt{-12} =$

$-\sqrt{25} \cdot \sqrt{2} + 2\sqrt{4} \cdot \sqrt{3} \cdot \sqrt{-1} =$
$\boxed{-5\sqrt{2} + 4\sqrt{3}\, i}$

5.7 Complex Numbers

1. Any number that can be expressed in the form $a + bi$, where a and b are real numbers and $i = \sqrt{-1}$, is called a <u>Complex #</u>. In the expression $a + bi$, a is called the <u>real part</u> and b is called the <u>imaginary part</u> of the complex number $a + bi$.

2. Write 1, -1, i, or $-i$ in the blank provided.

i = i	i^{14} = ________
i^{3} = $i^2 \cdot i = -i$	i^{66} = ________
i^{8} = $(i^4)^2 = (+i)^2 = 1$	i^{20} = ________
i^{100} = $(i^4)^{25} = 1$	i^{33} = ________
i^{37} = $(i^4)^9 \cdot i = i$	i^{43} = ________

$4/2 = 2$; $\frac{4}{25}$

$i = i$; $i^2 = -1$; $i^3 = -i$; $i^4 = 1$

3. $5 - 2i$ and $5 + 2i$ are called <u>Complex</u> <u>conjugate</u> of each other.

Simplify and express your answer in the form $a + bi$.

4. $(\sqrt{-36})(\sqrt{-100})$

$(\sqrt{36} \cdot \sqrt{-1})(\sqrt{100} \cdot \sqrt{-1})$

$6i \cdot 10i = 60i^2$

$= -1(60)$

$= -60 + 0i$

5. [5. / 17] $3\sqrt{18} - 2\sqrt{-24}$

$3\sqrt{9}\sqrt{2} - 2\sqrt{4}\sqrt{6}\sqrt{-1} =$

$9\sqrt{2} - 4\sqrt{6}\,i$

6. $\dfrac{(\sqrt{-6})(\sqrt{-12})}{\sqrt{-16}} = \dfrac{(\sqrt{6}\sqrt{-1})(\sqrt{4}\sqrt{3}\sqrt{-1})}{\sqrt{16}\sqrt{-1}}$

$\dfrac{\sqrt{6}\,i \cdot 2\sqrt{3}\,i}{4i} = \dfrac{i^2\, 2\sqrt{18}}{4i}$

$\dfrac{-1(2)\sqrt{18}}{4i} = \dfrac{-2\sqrt{9}\sqrt{2}}{4i}$

$\dfrac{-6\sqrt{2}}{4i} = \dfrac{-3\sqrt{2}}{2i} \cdot \dfrac{i}{i} = \dfrac{3i\sqrt{2}}{-2} = 0 + \dfrac{-3\sqrt{2}}{2}i$

7. $-\sqrt{50} + 2\sqrt{-12}$

$-\sqrt{25}\sqrt{2} + 2\sqrt{4}\sqrt{3}\sqrt{-1}$

$-5\sqrt{2} + 4\sqrt{3}\,i$

8. $(3 - 4i)^2 =$

$9 - 24i + 16i^2 =$

$9 - 24i - 16 =$

$\boxed{-7 - 24i}$

9. $(\sqrt{3} + 2i)(\sqrt{3} - i) =$

$\sqrt{9} - \sqrt{3}i + 2\sqrt{3}i - 2i^2 =$

$3 + \sqrt{3}i + 2 =$

$\boxed{5 + \sqrt{3}i}$

$\boxed{10.}$ $\boxed{41}$ $\left(-\frac{2}{3} + \frac{1}{9}i\right) - \left(\frac{3}{4} - \frac{1}{4}i\right) =$

$\frac{-2}{3} + \frac{1}{9}i - \frac{3}{4} + \frac{1}{4}i =$

$\boxed{\frac{-17}{12} + \frac{13}{36}i}$

11. $\frac{3}{i} \cdot \frac{i}{i} = \frac{3i}{i^2} = -3i =$

$\boxed{0 - 3i}$

12. $\frac{i}{3 + 4i} \cdot \frac{3 - 4i}{3 - 4i} =$

$\frac{3i - 4i^2}{9 - 16i^2} = \frac{3i + 4}{9 + 16} =$

$\frac{4 + 3i}{25} = \boxed{\frac{4}{25} + \frac{3}{25}i}$

$\boxed{13.}$ $\boxed{63}$ $\frac{4 + i}{2i} \cdot \frac{2i}{2i} =$

$\frac{8i + 2i^2}{4i^2} = \frac{8i - 2}{-4} =$

$\boxed{\frac{1}{2} - 2i}$

14. $-7 + 3i + (-4 + 6i) - (2 - 5i) =$

$-7 + 3i + {}^{-}4 + 6i - 2 + 5i =$

$\boxed{-13 + 14i}$

15. $(2 + 3i)\left(\frac{4 - 5i}{1 + i}\right) =$

$\frac{8 - 10i + 12i - 15i^2}{1 + i} =$

$\frac{23 + 2i}{1 + i} \cdot \frac{1 - i}{1 - i} =$

$\frac{23 - 23i + 2i - 2i^2}{1 - i^2} =$

$\frac{25 - 21i}{2} = \boxed{\frac{25}{2} - \frac{21}{2}i}$

8. $(3 - 4i)^2$

$(3-4i)(3-4i)$
$9-12i-12i+16i$
$9-24i+16(-1)$
$9-24i-16$
$-7-24i$

9. $(\sqrt{3} + 2i)(\sqrt{3} - i)$

$\sqrt{9} - i\sqrt{3} + 2i\sqrt{3} - 2i^2$
$3 + i\sqrt{3} - 2(-1)$
$5 + i\sqrt{3}$

10. (41) $\left(-\frac{2}{3} + \frac{1}{9}i\right) - \left(\frac{3}{4} - \frac{1}{4}i\right)$

11. $\frac{3}{i}$

12. $\frac{i}{3 + 4i}$

13. (63) $\frac{4 + i}{2i}$

14. $-7 + 3i + (-4 + 6i) - (2 - 5i)$

15. $(2 + 3i)\left(\frac{4 - 5i}{1 + i}\right)$

Chapter 5 Self-Test

Simplify each of the following. Remember to write your answer using only positive exponents. Assume that all variables represent positive real numbers.

1. $\left(\dfrac{x^6y^{\frac{1}{3}}}{49}\right)^{\frac{3}{2}}$

2. $\left(x^{\frac{2}{3}} - 3\right)\left(x^{\frac{2}{3}} + 4\right)$

3. $\sqrt[5]{-\dfrac{1}{243}}$

4. $\left(\sqrt[4]{3x^2y}\right)^8$

5. $\sqrt{x^5} \cdot \sqrt{x}$

6. $\sqrt[3]{54x^5y^2z^6}$

7. $\dfrac{\sqrt{-3x^2}}{\sqrt{-27x^4}}$

8. $\sqrt[3]{\dfrac{5xy^2}{6x^2y}} \cdot \sqrt[3]{\dfrac{x^6y}{36x^4y^2}}$

9. $3x\sqrt{75} - 2x\sqrt{27} - \sqrt{3x^2}$

10. $\dfrac{7}{\sqrt{5}} - 2\sqrt{5}$

11. $(7\sqrt{2} - 3)^2$

12. $\dfrac{5 - \sqrt{7}}{5 + \sqrt{7}}$

Solve the following radical equations.

13. $\sqrt{2x - 3} - 4 = 0$

14. $\sqrt{x + 3} - x = 1$

15. $\sqrt{2 - x} = 1 + \sqrt{3x + 1}$

Simplify and express your answer in the form $a + bi$.

16. $(\sqrt{5} - 3i)(\sqrt{5} + i)$

17. $\dfrac{2}{3 + 4i}$

CHAPTER

6 Linear Relations and Functions

6 · 1 The Cartesian Coordinate System

1. In the ordered pair (-3,4), the first coordinate or abscissa is -3 and the second coordinate or ordinate is 4.

2. The origin is the point at which the x-axis and the y-axis intersect in a rectangular coordinate system.

3. The first coordinate specifies movement in a horizontal direction; the second coordinate specifies movement in a vertical direction.

Match each ordered pair with its location in the Cartesian coordinate system.

a	4. (π, π)	(a) Quadrant I
b	5. $(-3, 15)$	(b) Quadrant II
e	6. $(3.21, 0)$	(c) Quadrant III
g	7. $(0, 0)$	(d) Quadrant IV

C H A P T E R

6 Linear Relations and Functions

6 · 1 The Cartesian Coordinate System

1. In the ordered pair (-3,4), the first coordinate or <u>abscissa</u> is -3 and the second coordinate or <u>ordinate</u> is 4.

2. The <u>origin</u> is the point at which the x-axis and the y-axis intersect in a <u>rectangular</u> coordinate system.

3. The <u>first</u> coordinate specifies movement in a horizontal direction; the <u>second</u> coordinate specifies movement in a vertical direction.

Match each ordered pair with its location in the Cartesian coordinate system.

a	4.	(π, π)	(a)	Quadrant I
b	5.	(-3, 15)	(b)	Quadrant II
e	6.	(3.21, 0)	(c)	Quadrant III
g	7.	(0, 0)	(d)	Quadrant IV

d	8. $(\sqrt{3}, -2.5)$	(e)	x-axis
c	9. $(-1/2, -5/6)$	(f)	y-axis
f	10. $(0, \sqrt{5})$	(g)	origin

11. Complete the table of values to find the ordered pairs which satisfy the equation $y = 4x^2 - 1$

x	y	
-2	15	$y = 4(-2)^2 - 1 = \underline{15}$
-1	3	$y = 4(-1)^2 - 1 = \underline{3}$
$\pm\frac{\sqrt{2}}{2}$	1	$(1) = 4x^2 - 1$ and solve for x: $1 = 4x^2 - 1$, $2 = 4x^2$, $x^2 = \frac{1}{2}$, $x = \pm\frac{\sqrt{2}}{2}$
$-.5$	0	$y = 4(-.5)^2 - 1 = \underline{0}$
2	15	$y = 4(2)^2 - 1 = \underline{15}$
$\pm\frac{1}{2}$	0	$(0) = 4x^2 - 1$ and solve for x: $0 = 4x^2 - 1$, $4x^2 = 1$, $x^2 = \frac{1}{4}$, $x = \pm\frac{1}{2}$

Construct a table of values for and graph each equation.

12. $y = 2x - 6$

x	y
0	-6
3	0
1	-4

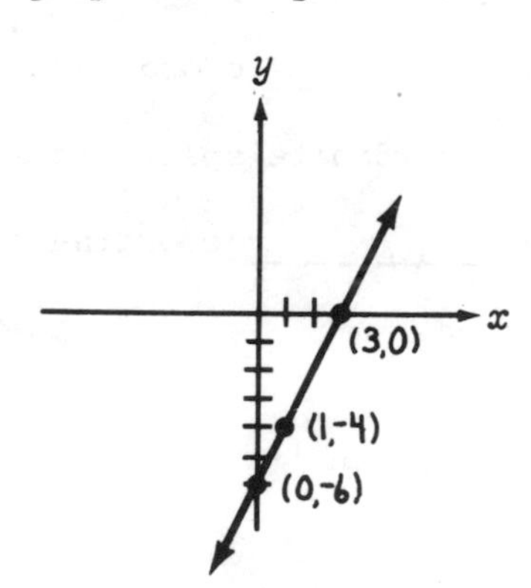

d 8. $(\sqrt{3}, -2.5)$ (e) x-axis

c 9. $(-1/2, -5/6)$ (f) y-axis

f 10. $(0, \sqrt{5})$ (g) origin

11. Complete the table of values to find the ordered pairs which satisfy the equation $y = 4x^2 - 1$

x	y	
-2	15	$y = 4(-2)^2 - 1 =$ 15
-1	3	$y = 4(-1)^2 - 1 =$ 3
$\pm \frac{\sqrt{2}}{2}$	1	$(1) = 4x^2 - 1$ and solve for x
-.5	0	$y = 4(-.5)^2 - 1 =$ 0
2	15	$y = 4(2)^2 - 1 =$ 15
$\pm \frac{1}{2}$	0	$(0) = 4x^2 - 1$ and solve for x

Construct a table of values for and graph each equation.

12. $y = 2x - 6$

x	y
2	-2
3	0
4	2

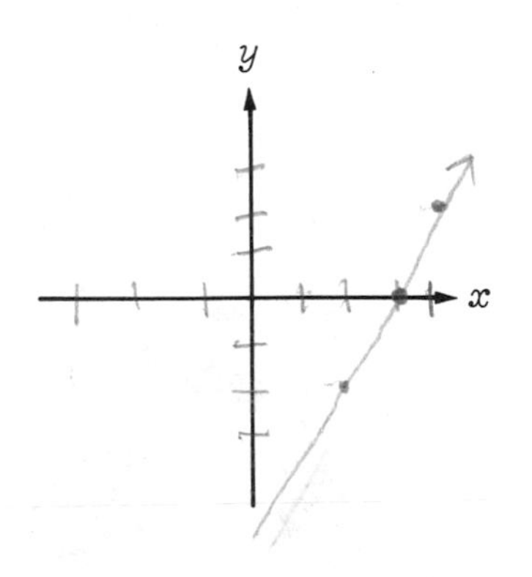

13. $y = x^2 - 2$

x	y
0	-2
1	-1
-1	-1

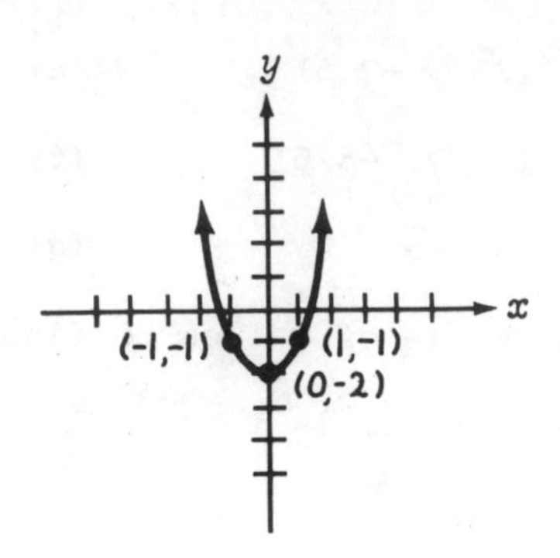

14. 15 $y = \sqrt{x} + 2$

x	y
0	2
4	4
9	5

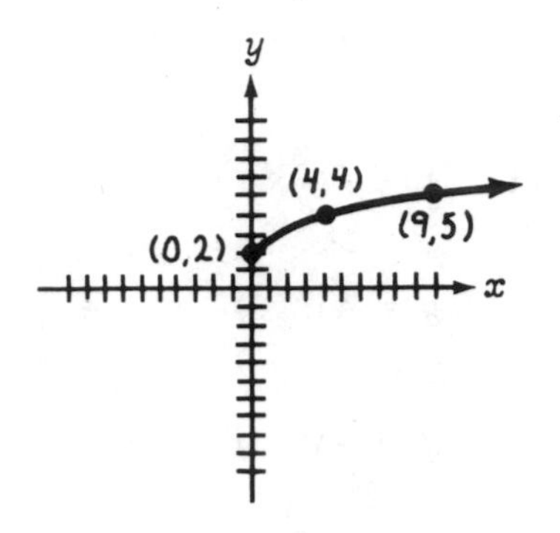

15. $x + 3y = y^2$ $x = y^2 - 3y$

x	y
0	0
-2	1
0	3

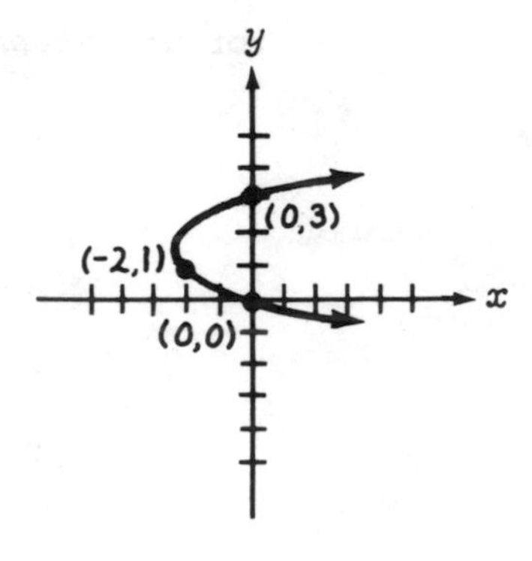

13. $y = x^2 - 2$

x	y
2	2
-2	2
0	-2
1	-1
-1	-1

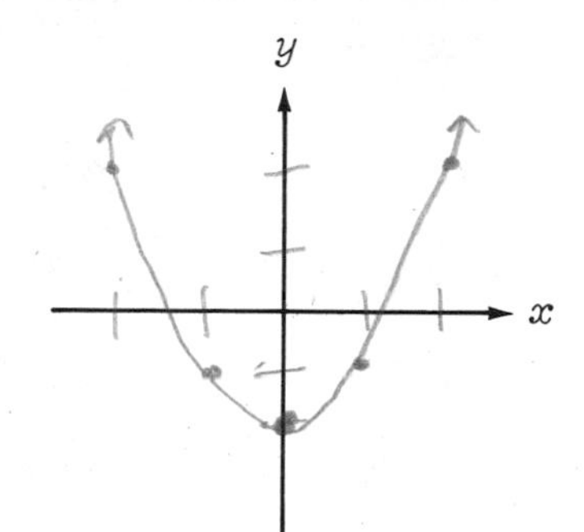

14. 15 $y = \sqrt{x} + 2$

x	y
0	2
4	4
9	5

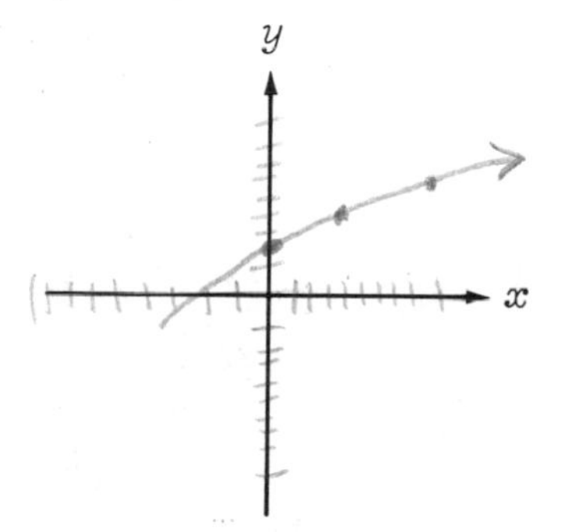

15. $x + 3y = y^2 - 3y$

$x = y(y-3)$

x	y
0	0
-2	1
0	3

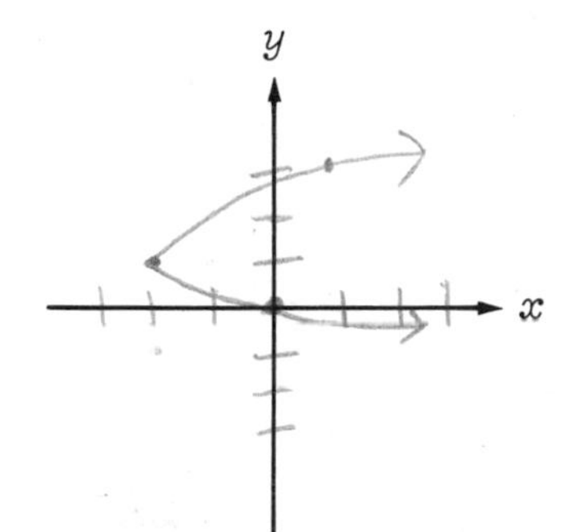

6·2 Relations and Functions

1. A set of ordered pairs is called a relation; the set of all first coordinates of the relation is called the domain; the set of all second coordinates of the relation is called the range.

2. A function is a relation such that each first coordinate is paired with one and only one second coordinate.

Write YES if a function; NO if not. Use the vertical-line test where appropriate.

yes 3. (0, 1), (1, 2), (2, 3), (3, 0)

no 4. (0, 1), (1, 2), (0, 2), (1, 3) x-values are paired to different y-values

no 5. yes 6. yes 7. no 8.

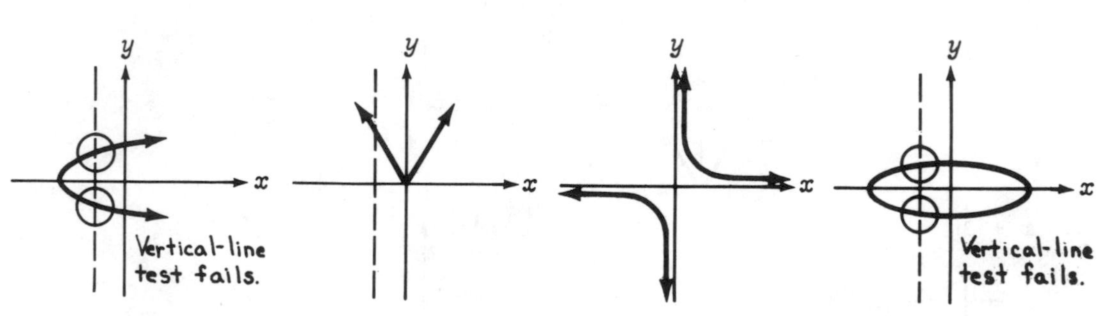

Specify the domain, D, and the range, R, of each of the following relations. Use set-builder notation for your answers.

9. {(2, 3), (4, 6), (8, 12)}

D = $\{x \mid x=2, x=4, x=8\}$

R = $\{y \mid y=3, y=6, y=12\}$

10. 7 $2x - y = 5$

D = $\{x \mid x$ is a real no.$\}$

R = $\{y \mid y$ is a real no.$\}$

6·2 Relations and Functions

1. A set of ordered pairs is called a relation; the set of all first coordinates of the relation is called the domain; the set of all second coordinates of the relation is called the range.

2. A function is a relation such that each first coordinate is paired with one and only one second coordinate.

Write YES if a function; NO if not. Use the vertical-line test where appropriate.

yes 3. (0, 1), (1, 2), (2, 3), (3, 0)

no 4. (0, 1), (1, 2), (0, 2), (1, 3)

no 5. yes 6. yes 7. no 8.

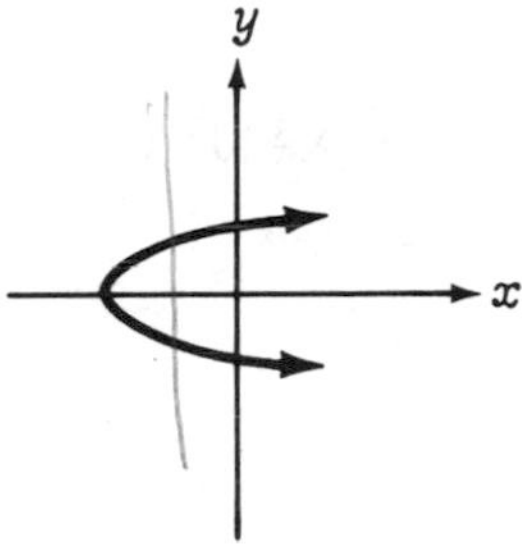

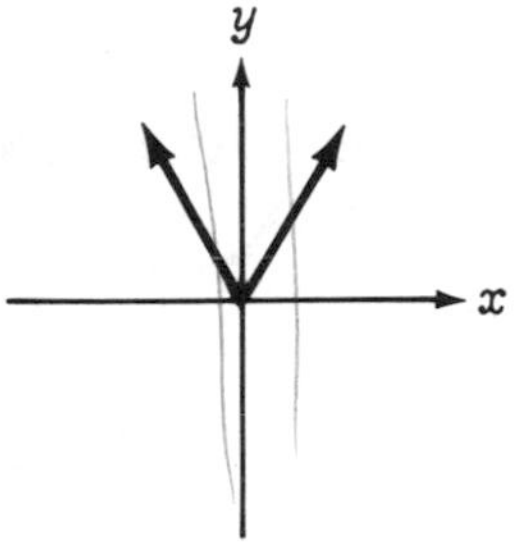

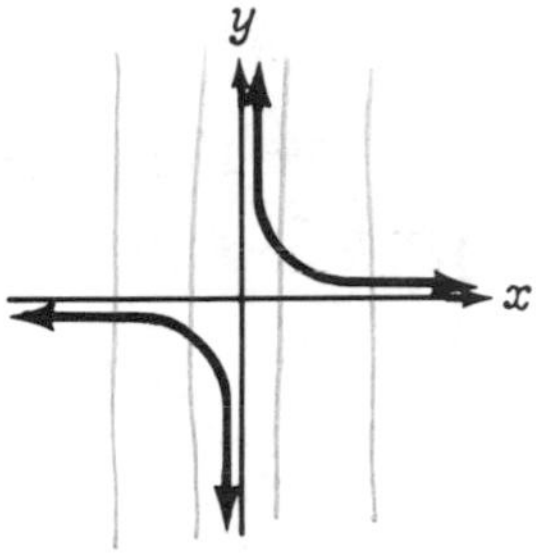

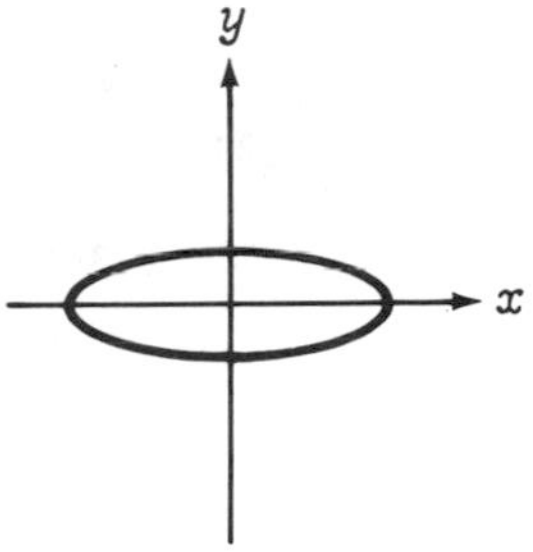

Specify the domain, D, and the range, R, of each of the following relations. Use set-builder notation for your answers.

9. {(2, 3), (4, 6), (8, 12)}

D = {x|x ≤ 2, x=4, x=8}

R = {y|y ≤ 3, y=6, y=12}

10. 7 $2x - y = 5$

D = {x|x is a real no}

R = {y|y is a real no}

11. $y = \sqrt{4 - x}$ 13

D = $\{x \mid x \leq 4\}$

R = $\{y \mid y \geq 0\}$

12. $y = \dfrac{7}{x - 3}$

D = $\{x \mid x \neq 3\}$

R = $\{y \mid y$ is a real no.$\}$

Given $f(x) = 2x^2 + x - 1$ and $g(x) = 4x - 2$, find:

13. $f(-3) = 2(-3)^2 + (-3) - 1 =$

$18 - 3 - 1 =$

$\boxed{14}$

14. $g(1/2) = 4(\frac{1}{2}) - 2 =$

$2 - 2 =$

$\boxed{0}$

15. $[f(-2)]^3 =$

$[2(-2)^2 + (-2) - 1]^3 =$

$(8 - 2 - 1)^3 = 5^3 =$

$\boxed{125}$

16. $f(x)/g(x) = \dfrac{2x^2 + x - 1}{4x - 2} =$

$\dfrac{(2x-1)(x+1)}{2(2x-1)} = \boxed{\dfrac{x+1}{2}}$

17. $f[g(2)] = f[4(2) - 2] = f[6]$

$f(6) = 2(6)^2 + (6) - 1 =$

$72 + 6 - 1 = \boxed{77}$

18. $f(x + 1) = 2(x+1)^2 + (x+1) - 1 =$

$2(x^2 + 2x + 1) + (x+1) - 1 =$

$2x^2 + 4x + 2 + x + 1 - 1 =$

$\boxed{2x^2 + 5x + 2}$

19. $g[f(2)] = g[2(2)^2 + 2 - 1] =$

$g(8 + 2 - 1) = g(9) =$

$g(9) = 4(9) - 2 =$

$36 - 2 = \boxed{34}$

20. $g(x^2 - 5) = 4(x^2 - 5) - 2$

$= 4x^2 - 20 - 2$

$= \boxed{4x^2 - 22}$

11. $y = \sqrt{4 - x}$ 13

D = $\{x \mid x \leq 4\}$

R = $\{y \mid y \geq 0\}$

12. $y = \dfrac{7}{x - 3}$

D = $\{x \mid x \neq 3\}$

R = $\{y \mid y$ is a real no$\}$

Given $f(x) = 2x^2 + x - 1$ and $g(x) = 4x - 2$, find:

13. $f(-3)$

$2(-3)^2 + -3 - 1$
$18 - 4$
14

14. $g(1/2)$ $4(\frac{1}{2}) - 2 =$
$2 - 2 = 0$

15. $[f(-2)]^3$

$-2^3 = -8$
$[2(-2)(-2) + -2 - 1]^3$
$(8 + -3)^3$
$(5)^3$
125

16. $f(x)/g(x)$

$\dfrac{2x^2 + x - 1}{4x - 2} = \dfrac{(2x - 1)(x + 1)}{2(2x - 1)}$

$\dfrac{x + 1}{2}$

17. $f[g(2)]$

$2x^2 + x - 1 \; [(4x - 2)2]$ 6
$2(6)^2 + 6 - 1$
$72 + 6 - 1$
77

18. $f(x + 1)$ $2(x+1)^2 + x + 1 - 1$
$2(x^2 + 2x + 1) + x + 0$
$2x^2 + 4x + 2 + x$
$2x^2 + 5x + 2$

19. $g[f(2)]$

$4x - 2 \; [2(2)^2 + 2 - 1]$
$8 = 8 + 1$
$9 = (9)4 - 2$
$36 - 2$
34

20. $g(x^2 - 5)$

$4(x^2 - 5) - 2$
$4x^2 - 20 - 2$
$4x^2 - 22$

21. $f(x - 1) - g(x - 1) =$

$[2(x-1)^2+(x-1)-1]-[4(x-1)-2]=$

$2x^2-4x+2+x-1-1-4x+4+2=$

$\boxed{2x^2-7x+6}$

22. $5f(-3) - 4g(2) =$

$5[2(-3)^2+(-3)-1]-4[4(2)-2]=$

$5(18-3-1)-4(8-2)=$

$5(14)-4(6)=$

$70-24= \boxed{46}$

6·3 Linear Functions

1. The graph of an equation of the form $Ax + By + C = 0$, where A, B, and C are real numbers and neither A nor B is zero, is a line.

2. The point, if any, where the graph of an equation intersects the x-axis is called the x-intercept. The point, if any, where the graph intersects the y-axis is called the y-intercept.

3. Given the equation, find its x-intercept and y-intercept.

	(a) $2x + y = 6$	(b) $y = -\pi$	(c) $3x - 2y = 9$	(d) $x = 4.5$
$(y=0)$ x-intercept	$2x+0=6$, $x=3$ $(3,0)$	none	$3x-2(0)=9$, $x=3$ $(3,0)$	$(4.5,0)$
$(x=0)$ y-intercept	$2(0)+y=6$, $y=6$ $(0,6)$	$(0,-\pi)$	$3(0)-2y=9$, $y=\frac{9}{2}$ $(0,-\frac{9}{2})$	none

4. Using the intercepts, graph each equation given in problem 3 above. Label your graph appropriately.

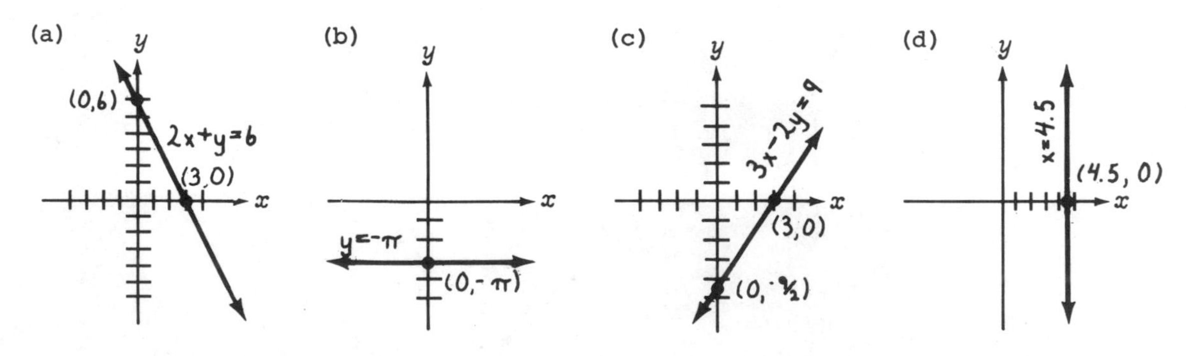

21. $f(x - 1) - g(x - 1)$

$[2(x-1)^2 + (x-1) - 1] - [4(x-1) - 2]$

$2x^2 - 4x + 2 + x - 1 - 1 - 4x + 4 + 2$

$-4x \quad -1$

$+x \quad -1$

$+4$

$+2$

$2x^2 - 7x + 6$

22. $5f(-3) - 4g(2)$

$5[2(-3)^2 + (-3) - 1] - 4[4(2) - 2]$

$8 - 2$

$5(18 - 4) \quad - 4(6)$

$5(14) \quad -24$

70

46

6·3 Linear Functions

1. The graph of an equation of the form $Ax + By + C = 0$, where A, B, and C are real numbers and neither A nor B is zero, is a line.

2. The point, if any, where the graph of an equation intersects the x-axis is called the x intercept. The point, if any, where the graph intersects the y-axis is called the y-intercept.

3. Given the equation, find its x-intercept and y-intercept.

	(a) $2x + y = 6$	(b) $y = -\pi$	(c) $3x - 2y = 9$	(d) $x = 4.5$
x-intercept	$2x + 0 = 6$ 3, 0	none	$3(3) - 2(0) = 9$ 3, 0	(4.5, 0)
y-intercept	$2(0) + 6 = 6$ (0, 6)	0, $-\pi$	$3(0) - \frac{2y}{-2} = -\frac{9}{2}$ 0, $-\frac{9}{2}$	none

4. Using the intercepts, graph each equation given in problem 3 above. Label your graph appropriately.

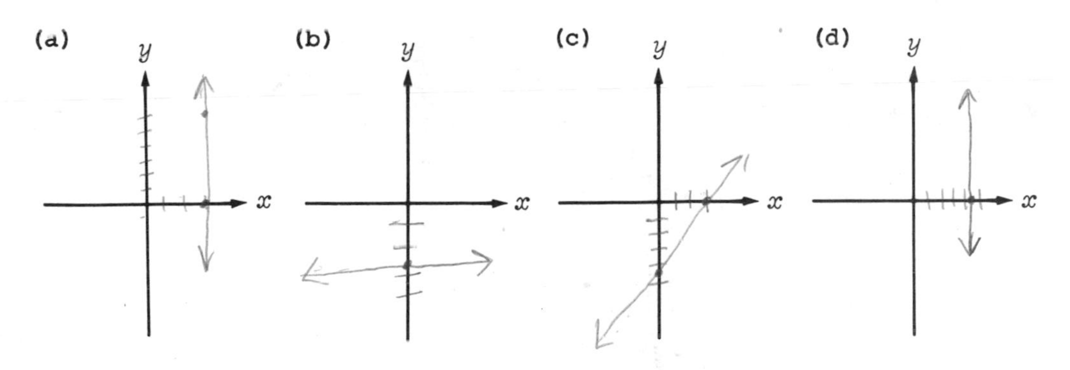

5. The slope of a line is a comparison of the vertical change to the horizontal change given by the formula $\frac{y_2 - y_1}{x_2 - x_1}$, where x_1 and y_1 are coordinates of one point on the line and x_2 and y_2 are coordinates of a second point on the line.

Find the slope of the line which passes through each of the following sets of ordered pairs.

6. $(-3, 5)$ and $(-1/2, -6)$

$$\frac{y_2-y_1}{x_2-x_1} = \frac{-6-5}{-\frac{1}{2}-{}^-3} = \frac{-11}{\frac{5}{2}} = \boxed{\frac{-22}{5}}$$

7. $(0, -3)$ and $(2, 1)$

$$\frac{y_2-y_1}{x_2-x_1} = \frac{1-{}^-3}{2-0} = \frac{4}{2} = \boxed{2}$$

8. $(4, 5)$ and $(4, -10)$

$$\frac{y_2-y_1}{x_2-x_1} = \frac{-10-5}{4-4} = \frac{-15}{0}$$

No slope

9. $(-2, 2/3)$ and $(-11, 3)$

$$\frac{y_2-y_1}{x_2-x_1} = \frac{3-\frac{2}{3}}{-11-{}^-2} = \frac{\frac{7}{3}}{-9} = \boxed{\frac{-7}{27}}$$

Given the slope and a point, graph the line.

10. $m = 2/3$, $(-3, 1)$

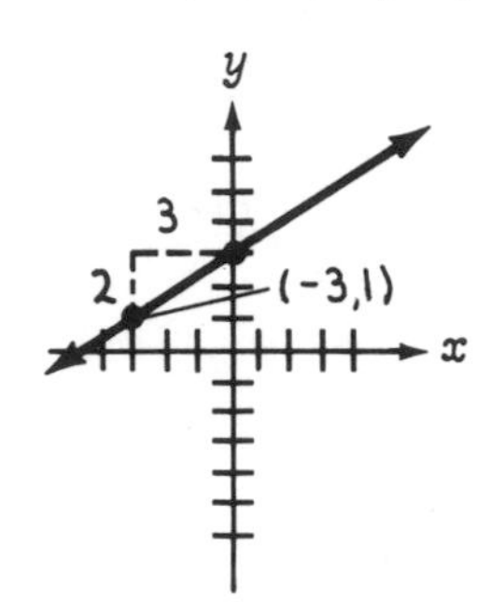

11. $m = -1/2$, $(4, -2)$

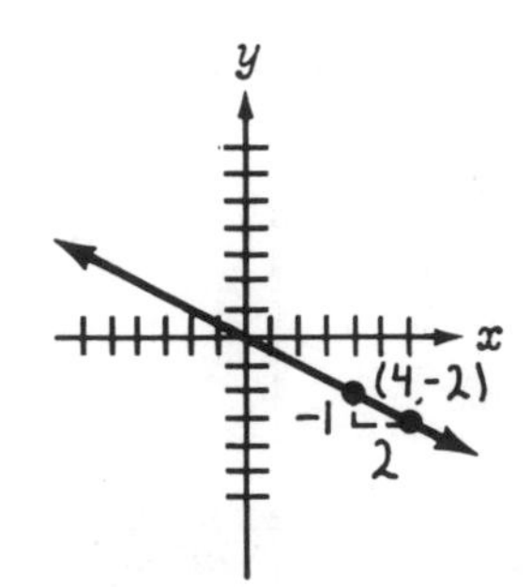

5. The slope of a line is a comparison of the vertical change to the horizontal change given by the formula $\frac{y_2 - y_1}{x_2 - x_1}$, where x_1 and y_1 are coordinates of one point on the line and x_2 and y_2 are coordinates of a second point on the line.

Find the slope of the line which passes through each of the following sets of ordered pairs.

6. (-3, 5) and (-1/2, -6)

$\frac{-6-5}{-1/2-(-3)} = \frac{-11}{\frac{5}{2}} = -11 \cdot \frac{2}{5}$

$-\frac{1}{2}+\frac{6}{2}$ $\frac{5}{2}$ $m = \frac{-22}{5}$

7. (0, -3) and (2, 1)

$\frac{1-(-3)}{2-0} = \frac{4}{2} = 2$

8. (4, 5) and (4, -10)

$\frac{-10-5}{4-4} = \frac{-15}{0}$

undefined

9. (-2, 2/3) and (-11, 3)

$\frac{3-2/3}{-11-(-2)} = \frac{\frac{7}{3}}{-9} = \frac{7}{3} \cdot \frac{1}{-9} = -\frac{7}{27}$

Given the slope and a point, graph the line.

10. $m = 2/3$, (-3, 1)

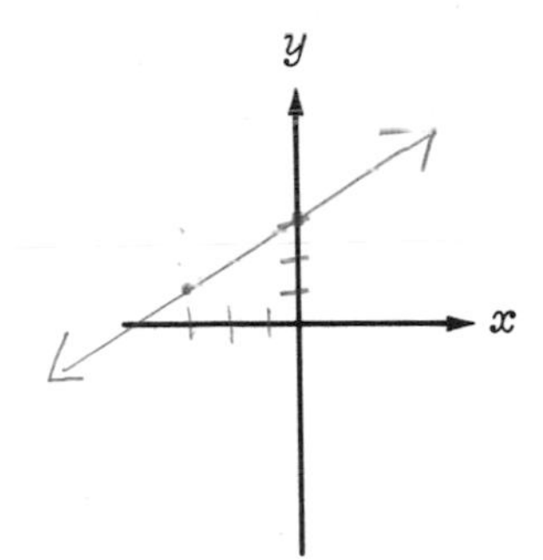

11. $m = -1/2$, (4, -2)

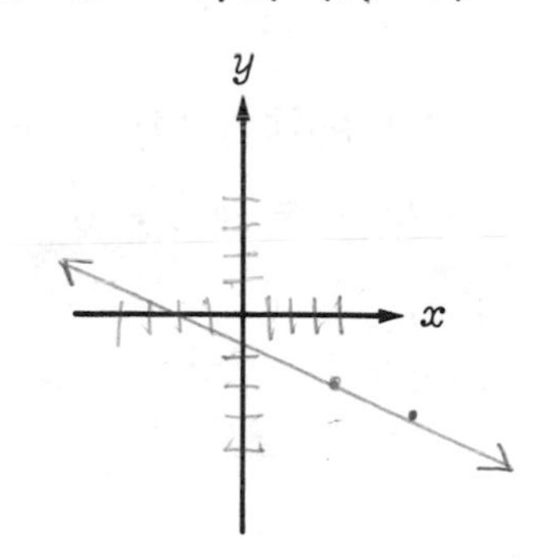

12. / 31 $m = 2, (3, -1)$

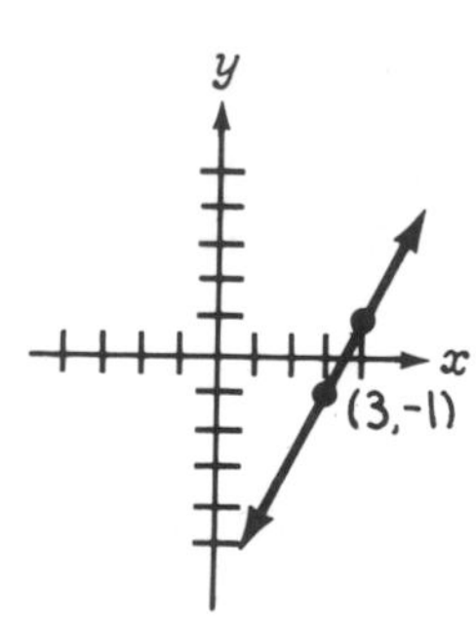

13. / 35 $m = 0, (5, 1)$

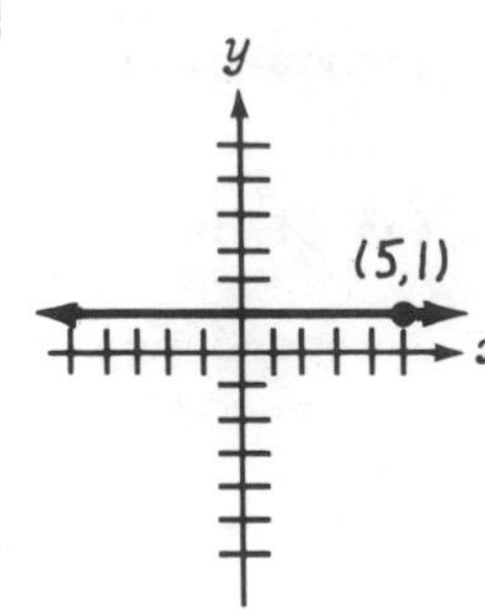

14. Find the slope of the line $2x + y = -3$ by use of the intercepts.

x-intercept: $(-\frac{3}{2}, 0)$ $y=0$ $2x+0=-3$ $x=\frac{-3}{2}$

y-intercept: $(0, -3)$ $x=0$ $2(0)+y=-3$ $y=-3$

Slope: $\frac{y_2-y_1}{x_2-x_1} = \frac{-3-0}{0-(-\frac{3}{2})} = \frac{-3}{\frac{3}{2}} = \boxed{-2}$

6·4 Equations of Lines

1. If a line passes through at least one known point, (x_1, y_1), and the slope m of the line is known, then the equation of the line is found by the formula $y-y_1=m(x-x_1)$. Since a point and the slope are known, this formula is called the point-slope form of the equation of a line.

Given the slope and a point, find the equation of each line. Express answer in the form $Ax + By + C = 0$.

2. / 21 $m = 9/2$, $(3, 8)$ (x_1, y_1)

$y-y_1=m(x-x_1)$
$y-8=\frac{9}{2}(x-3)$
$y-8=\frac{9}{2}x-\frac{27}{2}$
$\boxed{\frac{9}{2}x-y-\frac{11}{2}=0}$ OR $9x-2y-11=0$

3. $m = 2/(-3)$, $(0, -1)$

$y-y_1=m(x-x_1)$
$y+1=\frac{2}{-3}(x-0)$
$y+1=-\frac{2}{3}x$
$\boxed{\frac{2}{3}x+y+1=0}$ OR $2x+3y+3=0$

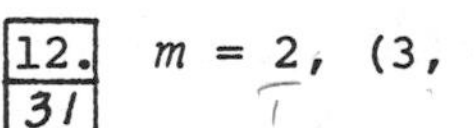

12. (31) $m = 2$, $(3, -1)$

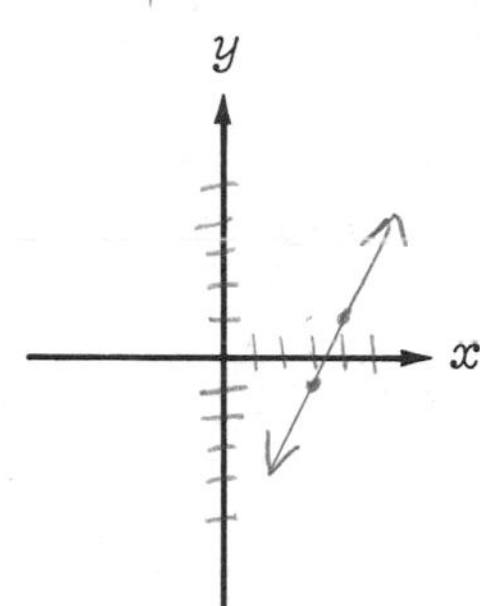

13. (35) $m = 0$, $(5, 1)$

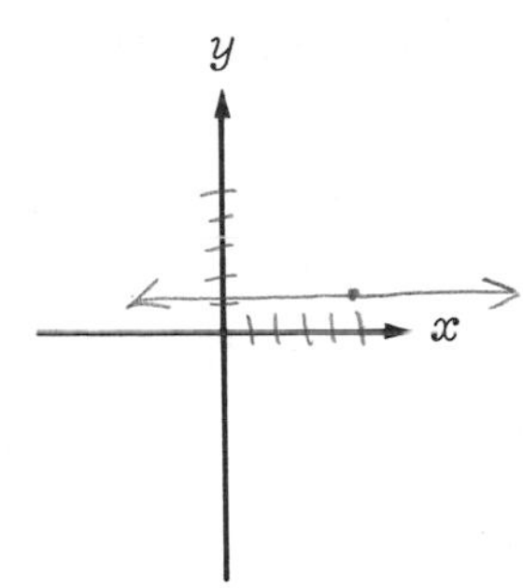

14. Find the slope of the line $2x + y = -3$ by use of the intercepts.

x-intercept: $-\frac{3}{2}, 0$ $\qquad 2x + 0 = -3$, $\frac{2x}{2} = \frac{-3}{2}$

y-intercept: $0, -3$ $\qquad 2(0) + y = -3$, $-3 = -3$

Slope: $\frac{-3 - 0}{0 - (-\frac{3}{2})} = \frac{-3}{\frac{3}{2}} = -\frac{3}{1} \cdot \frac{2}{3} = -\frac{6}{3} = -2$

6·4 Equations of Lines

1. If a line passes through at least one known point, (x_1, y_1), and the slope m of the line is known, then the equation of the line is found by the formula $y - y_1 = m(x - x_1)$. Since a point and the slope are known, this formula is called the point-slope form of the equation of a line.

Given the slope and a point, find the equation of each line. Express answer in the form $Ax + By + C = 0$.

2. (21) $m = 9/2$, $(3, 8)$

$y - 8 = \frac{9}{2}(x - 3)$

$y - 8 = \frac{9}{2}x - \frac{27}{2}$

$\frac{27}{2} - 8 = \frac{9}{2}x - y$

$\frac{27}{2} - \frac{16}{2} = \frac{9}{2}x - y$

$\frac{11}{2} = \frac{9}{2}x - y$ or $\frac{9}{2}x - y - \frac{11}{2} = 0$

$9x - 2y - 11 = 0$

3. $m = 2/(-3)$, $(0, -1)$

$y + 1 = \frac{2}{-3}(x - 0)$

$y + 1 = \frac{2}{-3}x$

$= \frac{2}{-3}x - y - 1$

$= 2x + 3y + 3 = 0$

Given that a line passes through the two given points, (a) find the slope and (b) find the equation of the line.

4. [29] $(-8, 6)$ and $(4, -3)$ (x_1, y_1; x_2, y_2)

(a) slope: $\frac{y_2-y_1}{x_2-x_1} = \frac{-3-6}{4-(-8)} = \frac{-9}{12} = \boxed{\frac{-3}{4}}$

(b) equation of the line:

$y - y_1 = m(x - x_1)$

$y - 6 = -\frac{3}{4}(x - -8)$

$y - 6 = \frac{-3}{4}x - 6$

$\boxed{\frac{3}{4}x + y = 0}$ OR $3x + 4y = 0$

5. $(-1/2, 3)$ and $(-5/2, -6)$ (x_1, y_1; x_2, y_2)

(a) slope: $\frac{y_2-y_1}{x_2-x_1} = \frac{-6-3}{-\frac{5}{2}-\frac{-1}{2}} = \frac{-9}{-2} = \boxed{\frac{9}{2}}$

(b) equation of the line:

$y - y_1 = m(x - x_1)$

$y - 3 = \frac{9}{2}(x - \frac{-1}{2})$

$y - 3 = \frac{9}{2}x + \frac{9}{4}$

$\boxed{\frac{9}{2}x - y + \frac{21}{4} = 0}$ OR

$18x - 4y + 21 = 0$

6. If a line has a known y-intercept, $(0, b)$, and a known slope m, then the equation of the line is found by the formula $y = mx + b$.

Since the slope and the y-intercept are known, this is called the slope-intercept form of the equation of a line.

Find the equation of each line with the given slope and the given y-intercept. Express your answer in the form $y = mx + b$.

7. [5] $m = -1/2$, y-intercept at $(0, -3)$

$y = mx + b$

$\boxed{y = \frac{-1}{2}x - 3}$

8. $m = 0$, y-intercept at $(0, 0)$

$y = mx + b$

$y = 0x + 0$

$\boxed{y = 0}$

NOTE: The above problems, 7 and 8, could be worked, if desired, as point-slope problems.

Given that a line passes through the two given points, (a) find the slope and (b) find the equation of the line.

4. **29** (-8, 6) and (4, -3)

$\frac{-3-6}{4+8} = \frac{-9}{12} = -\frac{3}{4}$

(a) slope: $-\frac{3}{4}$

(b) equation of the line:

$y-6 = -\frac{3}{4}(x-8)$

$y-6 = -\frac{3}{4}x + 6$

$0 = -\frac{3}{4}x + 6 - y + 6$

$0 = 3x - 24 + 4y - 24$

$0 = 3x + 4y$

5. (-1/2, 3) and (-5/2, -6)

(a) slope: $\frac{-6-3}{-5/2+\frac{1}{2}} = \frac{-9}{-\frac{4}{2}} = \frac{9}{2}$

(b) equation of the line:

$y-3 = \frac{9}{2}(x+\frac{1}{2})$

$y-3 = \frac{9}{2}x + \frac{9}{4}$

$0 = \frac{9}{2}x + \frac{9}{4} - y + 3$

$4(\frac{9}{2}x) + 4(\frac{9}{4}) + 4(-y) + 4(3)$

$0 = 18x + 9 - 4y + 12$

$0 = 18x - 4y + 21$

6. If a line has a known y-intercept, $(0, b)$, and a known slope m, then the equation of the line is found by the formula $y = mx + b$. Since the slope and the y-intercept are known, this is called the slope-intercept form of the equation of a line.

Find the equation of each line with the given slope and the given y-intercept. Express your answer in the form $y = mx + b$.

7. **5** $m = -1/2$, y-intercept at (0, -3)

$y = -\frac{1}{2}x - 3$

8. $m = 0$, y-intercept at (0, 0)

$y = 0x + 0$

$y = 0$

NOTE: The above problems, 7 and 8, could be worked, if desired, as point-slope problems.

Find the slope and the y-intercept of the following lines.

9. $y = (-2/3)x - 3$

$m = \frac{-2}{3}$

y-intercept: $(0, -3)$

10. (15) $y = \frac{2x - 4}{7} = \frac{2}{7}x - \frac{4}{7}$

$m = \frac{2}{7}$

y-intercept: $(0, -\frac{4}{7})$

11. $-2x + 3y = -9$

$3y = 2x - 9$

$y = \frac{2}{3}x - 3$

$m = \frac{2}{3}$

y-intercept: $(0, -3)$

12. $x = -8$ (vertical line)

No slope

No y-intercept

13. Two lines are parallel if they have equal slopes; they are perpendicular if their slopes are negative reciprocals.

Find the equation of each line passing through the given point which is (a) parallel to the given line and (b) perpendicular to the given line.

14. $(2, 6)$; $3x + y = 5$

(a) $m = -3$

$y - 6 = -3(x - 2)$

$y - 6 = -3x + 6$

$y = -3x + 12$

(b) $m = +\frac{1}{3}$

$y - 6 = \frac{1}{3}(x - 2)$

$y - 6 = \frac{1}{3}x - \frac{2}{3}$

$y = \frac{1}{3}x + \frac{16}{3}$

Find the slope and the y-intercept of the following lines.

9. $y = (-2/3)x - 3$

$m = \underline{-\frac{2}{3}}$

y-intercept: $(0 , -3)$

10. [15] $y = \dfrac{2x - 4}{7}$

$y = \frac{2}{7}x - \frac{4}{7}$

$m = \frac{2}{7}$

$y = (0, -\frac{4}{7})$

11. $-2x + 3y = -9$

$\frac{3y}{3} = \frac{2x}{3} - \frac{9}{3}$

$y = \frac{2}{3}x - 3$

$m = \frac{2}{3}$

$y = -3$

$(0, -3)$ y-intercept

12. $x = -8$

0

no y intercept

13. Two lines are <u>parallel</u> if they have equal slopes; they are <u>perpendicular</u> if their slopes are negative reciprocals.

Find the equation of each line passing through the given point which is (a) parallel to the given line and (b) perpendicular to the given line.

$3x + y - 5 = 0$

14. $(2, 6)$; $3x + y = 5$

(a) $y - 6 = -3(x - 2)$

$y - 6 = -3x + 6$

$y = -3x + 12$

(b) $m = \frac{1}{3}$

$y - y_1 = \frac{1}{3}(x - x_1)$

$y - 6 = \frac{1}{3}(x - 2)$

$y - 6 = \frac{1}{3}x - \frac{2}{3}$

$y = \frac{1}{3}x - \frac{2}{3} + 6$

$y = \frac{1}{3}x - \frac{2}{3} + \frac{18}{3}$

$y = \frac{1}{3}x - \frac{16}{3}$

15. $(1/4, -3/5)$; $y = \frac{6x - 13}{8}$

(a) $m = \frac{3}{4}$

$y - \frac{-3}{5} = \frac{3}{4}(x - \frac{1}{4})$

$y + \frac{3}{5} = \frac{3}{4}x - \frac{3}{16}$

$\boxed{y = \frac{3}{4}x - \frac{63}{80}}$

(b) $m = -\frac{4}{3}$

$y - \frac{-3}{5} = -\frac{4}{3}(x - \frac{1}{4})$

$y + \frac{3}{5} = -\frac{4}{3}x + \frac{1}{3}$

$\boxed{y = -\frac{4}{3}x - \frac{4}{15}}$

16. $(2.5, -1)$; y-axis

(a) $\boxed{x = 2.5}$

(b) $m = 0$

$y - -1 = 0(x - 2.5)$

$y + 1 = 0$

$\boxed{y = -1}$

17. $(-1, 2)$; x-axis

(a) $m = 0$

$y - 2 = 0(x - -1)$

$y - 2 = 0$

$\boxed{y = 2}$

(b) $\boxed{x = -1}$

18. / 61 Show that the following four points are vertices of a rectangle: $A = (-2, -5)$, $B = (4, -2)$, $C = (2, 2)$, and $D = (-4, -1)$.

We must show that: (1) opposite sides are parallel and (2) adjacent sides are perpendicular.

Slope of $\overline{AB} = \frac{-2 - -5}{4 - -2} = \frac{3}{6} = \frac{1}{2}$

Slope of $\overline{BC}$ $\frac{2 - -2}{2 - 4} = \frac{4}{-2} = -2$

Slope of $\overline{CD}$ $\frac{-1 - 2}{-4 - 2} = \frac{-3}{-6} = \frac{1}{2}$

Slope of $\overline{AD} = \frac{-1 - -5}{-4 - -2} = \frac{4}{-2} = -2$

$\overline{AB} \parallel \overline{CD}$, $\overline{BC} \parallel \overline{AD}$ — Parallel because they have equal slopes.

$\overline{AB} \perp \overline{BC}$, $\overline{BC} \perp \overline{CD}$, $\overline{CD} \perp \overline{AD}$, $\overline{AD} \perp \overline{AB}$ — Perpendicular because slopes are negative reciprocals.

Therefore, ABCD is a rectangle.

15. $(1/4, -3/5)$; $y = \dfrac{6x - 13}{8}$ $y = \frac{6x}{8} - \frac{13}{8}$

(a) $y = \frac{3}{4}$

$y - \frac{3}{5} = \frac{3}{4}(x - \frac{1}{4})$

$y = \frac{3}{4}x - \frac{3}{16} + \frac{3}{5}$

$y = \frac{3}{4}x - \frac{63}{80}$

(b) $m = -\frac{4}{3}$

$y - (-\frac{3}{5}) = -\frac{4}{3}(x - \frac{1}{4})$

$y + \frac{3}{5} = -\frac{4}{3}x + \frac{4}{12}$

$y = -\frac{4}{3}x + \frac{1}{3} - \frac{3}{5}$

$y = -\frac{4}{3}x + \frac{5}{15} - \frac{9}{15}$

$y = -\frac{4}{3}x - \frac{4}{15}$

16. $(2.5, -1)$; y-axis

(a) $x = 2.5$

(b)

$m = 0$

$y - (-1) = 0(x - 2.5)$

$y + 1 = 0$

$y = -1$

17. $(-1, 2)$; x-axis

(a) $m = 0$

$y - 2 = 0(x - (-1))$

$y = 2$

(b) $x = -1$

18. / 61 Show that the following four points are vertices of a rectangle: $A = (-2, -5)$, $B = (4, -2)$, $C = (2, 2)$, and $D = (-4, -1)$.

slope $\overline{AB} = \frac{-2 - (-5)}{4 - (-2)} = \frac{3}{6} = \frac{1}{2}$

slope $\overline{BC} = \frac{2 - (-2)}{2 - 4} = \frac{4}{-2} = -2$

slope $\overline{CD} = \frac{-1 - 2}{-4 - 2} = \frac{-3}{-6} = \frac{1}{2}$

slope $\overline{AD} = \frac{-1 - (-5)}{-4 - (-2)} = \frac{4}{-2} = -2$

$\overline{AB} \parallel \overline{CD}$, $\overline{BC} \parallel \overline{AD}$ } equal m's

$\overline{AB} \perp \overline{BC}$ perpendicular

$BC \perp CD$ "

$CD \perp AD$ "

negative reciprocals

6 · 5 Distance Formula

1. The distance between any two points, (x_1, y_1) and (x_2, y_2), is found by the formula $\sqrt{(x_2-x_1)^2+(y_2-y_1)^2}$.

Determine the length of each segment between the given points.

2.
9

(0, 3) and (-2, -6)

$$d = \sqrt{(x_2-x_1)^2+(y_2-y_1)^2} =$$

$$\sqrt{(-2-0)^2+(-6-3)^2} =$$

$$\boxed{d = \sqrt{85} \doteq 9.22}$$

3.
14

(-4/3, 1) and (-1/2, -1/4)

$$d = \sqrt{(x_2-x_1)^2+(y_2-y_1)^2} =$$

$$\sqrt{\left(-\tfrac{1}{2}+\tfrac{4}{3}\right)^2+\left(-\tfrac{1}{4}-1\right)^2} =$$

$$\sqrt{\frac{25}{36}+\frac{25}{16}} = \sqrt{\frac{100}{576}} \cdot \sqrt{13} =$$

$$\boxed{\frac{5}{12}\sqrt{13} \doteq 1.502}$$

4. (3/4, 0) and (0, -1/4)

$$d = \sqrt{(x_2-x_1)^2+(y_2-y_1)^2} =$$

$$\sqrt{\left(0-\tfrac{3}{4}\right)^2+\left(-\tfrac{1}{4}-0\right)^2} =$$

$$\sqrt{\frac{9}{16}+\frac{1}{16}} =$$

$$\boxed{\frac{1}{4}\sqrt{10} \doteq .791}$$

6 · 5 Distance Formula

1. The distance between any two points, (x_1, y_1) and (x_2, y_2), is found by the formula $\sqrt{(x_2-x_1)^2+(y_2-y_1)^2}$.

Determine the length of each segment between the given points.

2. | 9 | (0, 3) and (-2, -6)

$$\sqrt{(-2-0)^2+(-6-3)^2}$$
$$\sqrt{(-2)^2+(-9)^2}$$
$$= \sqrt{4+81}$$
$$d = \sqrt{85} = 9.22$$

3. | 14 | (-4/3, 1) and (-1/2, -1/4)

4. (3/4, 0) and (0, -1/4)

5. $(2a, 3)$ and $(-a, 5)$

$$d = \sqrt{(x_2 - x_1)^2 + (y_2 - y_1)^2} =$$
$$\sqrt{(-a - 2a)^2 + (5 - 3)^2} =$$
$$\sqrt{(-3a)^2 + 2^2} =$$
$$\boxed{d = \sqrt{9a^2 + 4}}$$

6. | 21

Use the distance formula and the Pythagorean Theorem to show that the following points are the vertices of a right triangle:

$A = (-1, 6)$, $B = (1, 2)$, and $C = (-5, -1)$

(1)
$$d(\overline{AB}) = \sqrt{(-1-1)^2 + (6-2)^2} = \sqrt{4+16} = \sqrt{20}$$
$$d(\overline{BC}) = \sqrt{(1+5)^2 + (2+1)^2} = \sqrt{36+9} = \sqrt{45}$$
$$d(\overline{AC}) = \sqrt{(-1+5)^2 + (6+1)^2} = \sqrt{16+49} = \sqrt{65}$$

(2) $\overline{AC}$ is the hypotenuse because it is the longest side if ABC is a right triangle.

(3) Pythagorean Theorem:
$$a^2 + b^2 = c^2$$
$$(\sqrt{20})^2 + (\sqrt{45})^2 \stackrel{?}{=} (\sqrt{65})^2$$
$$20 + 45 = 65$$
$$65 \stackrel{\checkmark}{=} 65$$

(4) Therefore, ABC is a right triangle.

5. $(2a, 3)$ and $(-a, 5)$

6.
Z1

Use the distance formula and the Pythagorean Theorem to show that the following points are the vertices of a right triangle:

$A = (-1, 6)$, $B = (1, 2)$, and $C = (-5, -1)$

$D(AB) = \sqrt{(1-(-1))^2 + (2-6)^2}$

$(2)^2 + (-4)^2$

$D(AB) = \sqrt{4 + 16}$

$D(AB) = \sqrt{20}$

$D(AB) = 2\sqrt{5}$

$D(BC) = \sqrt{(-5-1)^2 + (-1-2)^2}$

$\sqrt{(-6)^2 + (-3)^2}$

$\sqrt{36 + 9}$

$D(BC) = \sqrt{45}$

$D(AC) = \sqrt{(-5-(-1))^2 + (-1-6)}$

$\sqrt{(-4)^2 + (-7)^2}$

$\sqrt{16 + 49}$

$D(AC) = \sqrt{65}$

$(\sqrt{65})^2 = (\sqrt{20})^2 + (\sqrt{45})^2$

$65 = 65$

6 · 6 Linear Inequalities in Two Variables

1. Graph $y < x + 5$

(a) Graph the boundary line $y = x + 5$

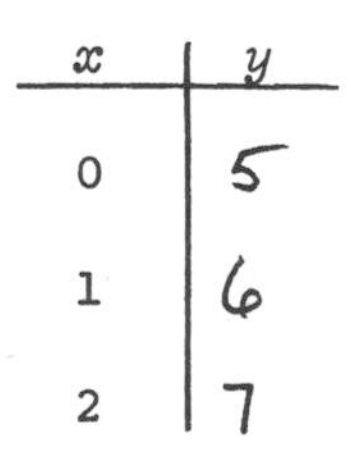

x	y
0	5
1	6
2	7

This will be a dashed line because the original inequality contains <.

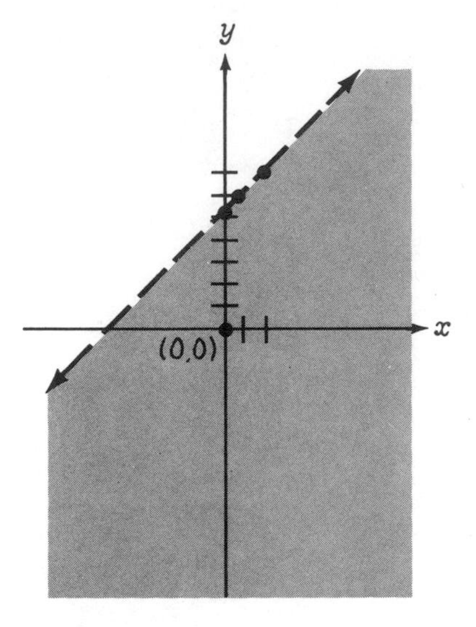

(b) Decide which half-plane should be shaded.

Pick a point: (0, 0)

Substitute into the original inequality: $y < x + 5$

$\underline{0} < \underline{0} + 5$

$\underline{0} < \underline{5}$ TRUE

Since (0, 0) gives a true result, shade the half-plane containing (0, 0).

Check: Pick another point: (1, 10)

Substitute: $y < x + 5$

$\underline{10} < \underline{1} + 5$

$\underline{10} < \underline{6}$ FALSE

Since (1, 10) gives a false result, do not shade the half-plane containing (1, 10).

6·6 Linear Inequalities in Two Variables

1. Graph $y < x + 5$

(a) Graph the boundary line $y = x + 5$

x	y
0	5
1	6
2	7

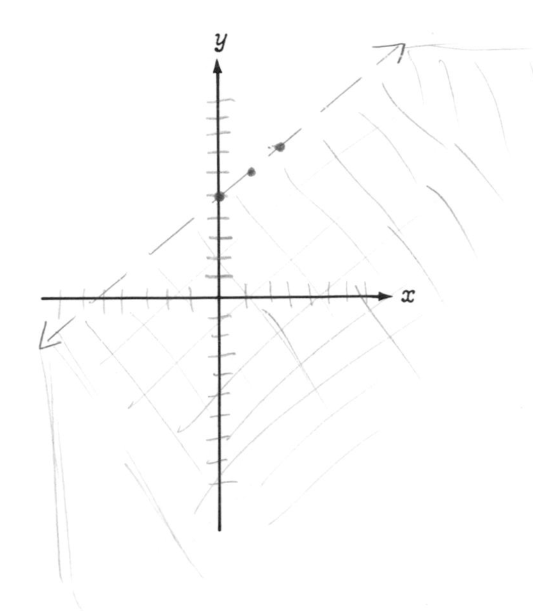

This will be a dashed line because the original inequality contains <.

(b) Decide which half-plane should be shaded.

Pick a point: (0, 0)

Substitute into the original inequality: $y < x + 5$

$\underline{0} < \underline{0} + 5$

$\underline{0} < \underline{5}$ TRUE

Since (0, 0) gives a true result, shade the half-plane containing (0, 0).

Check: Pick another point: (1, 10)

Substitute: $y < x + 5$

$\underline{10} < \underline{1} + 5$

$\underline{10} < \underline{6}$ FALSE

Since (1, 10) gives a false result, do not shade the half-plane containing (1, 10).

2. (9) Graph $5x > -3y + 7$

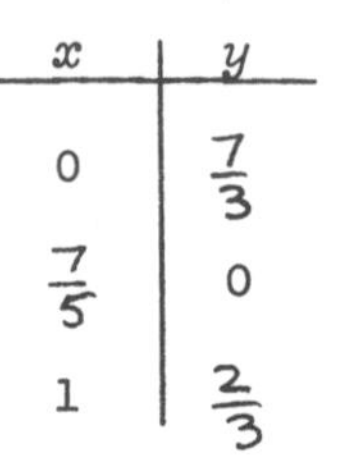

x	y
0	$\frac{7}{3}$
$\frac{7}{5}$	0
1	$\frac{2}{3}$

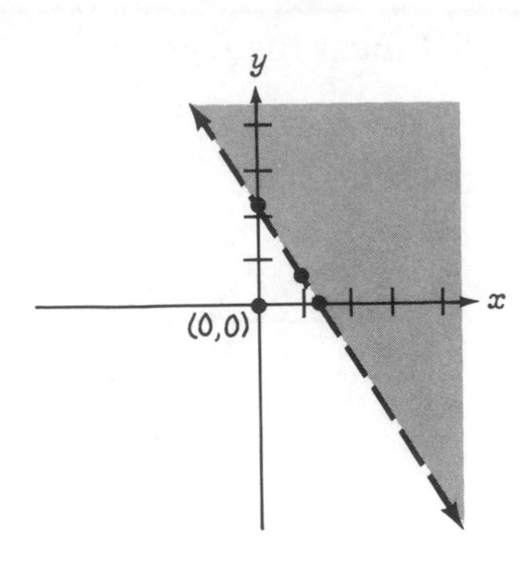

Check (0,0): $5(0) > (-3)(0) + 7$

$0 > 7$ False

3. Graph $y \leq -2x + 3$

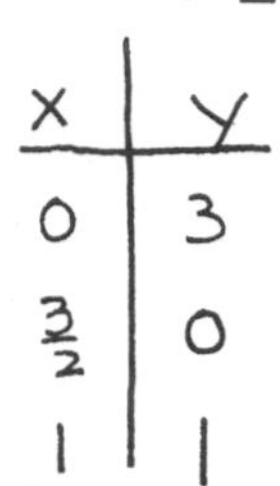

X	Y
0	3
$\frac{3}{2}$	0
1	1

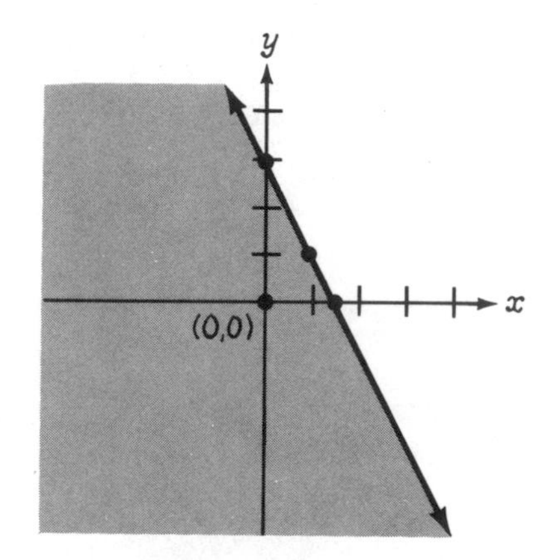

Check (0,0): $0 \leq -2(0)+3$

$0 \leq 3$ True

Solve the following system of inequalities by graphing the inequalities on the same Cartesian coordinate system. Examine where the shaded areas overlap to find the set of points that satisfies the system of inequalities.

4. (19) $2x + y < 5$ and $3x - y < 1$

X	Y
0	5
$\frac{5}{2}$	0
1	3

Test (0,0):
$2(0)+(0)<5$
$0<5$ True

X	Y
0	-1
$\frac{1}{3}$	0
1	2

Test (0,0):
$3(0)-(0)<1$
$0<1$ True

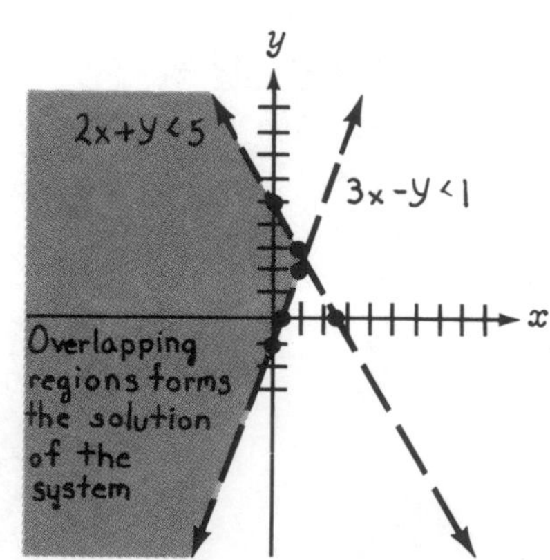

2. **9** Graph $5x > -3y + 7$

x	y
0	$\frac{7}{3}$
$\frac{7}{5}$	0
1	$\frac{2}{3}$

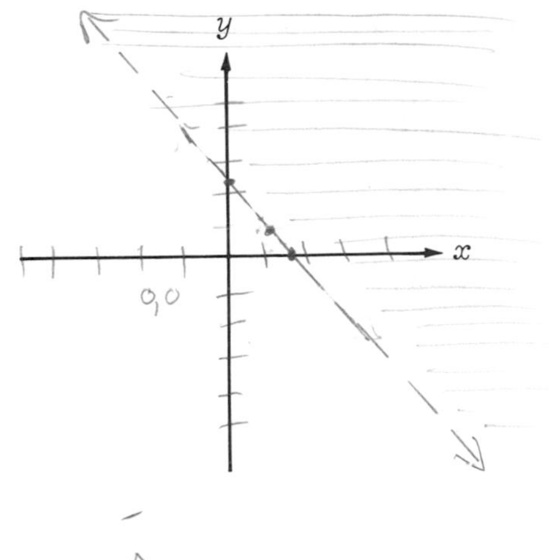

$5(0) > -3(0) + 7$
$0 > 0 + 7$
false

3. Graph $y \leq -2x + 3$

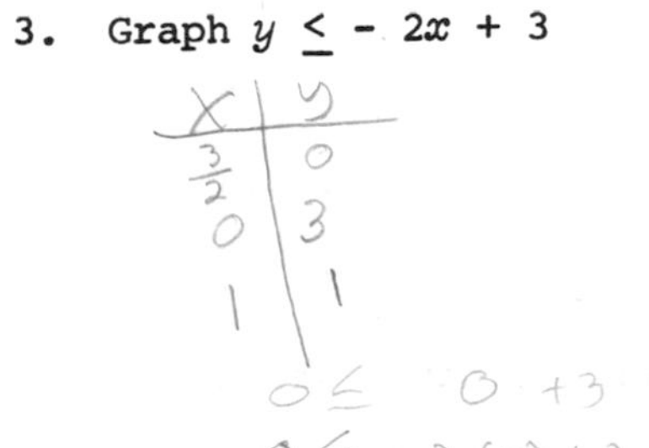

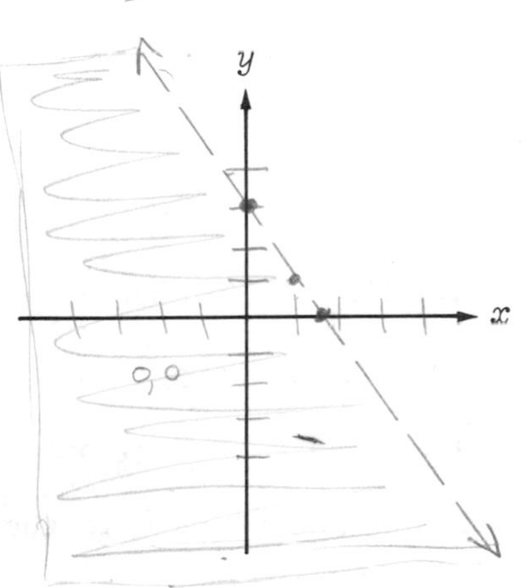

Solve the following system of inequalities by graphing the inequalities on the same Cartesian coordinate system. Examine where the shaded areas overlap to find the set of points that satisfies the system of inequalities.

4. **19** $2x + y < 5$ and $3x - y < 1$

X	y
0	5
$\frac{5}{2}$	0
1	3

X	y
0	-1
$\frac{1}{3}$	0
1	2

$2(0) + 0 < 5$ true
$3(0) - 0 < 1$ true

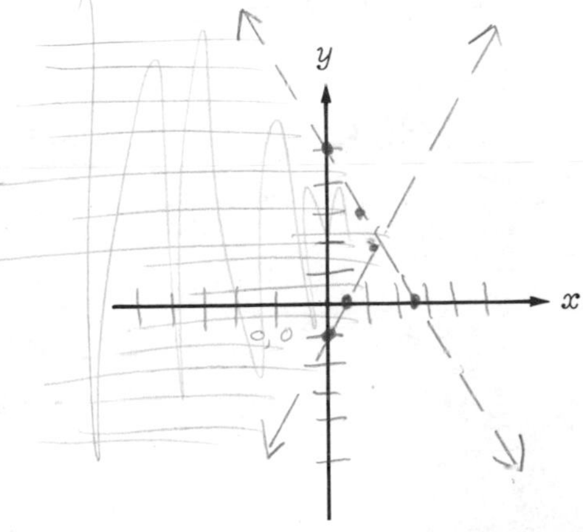

5. $x - y < -3$ and $2x + y \leq 5$ and $x \geq 1$

x	y
0	3
-3	0

Check (0,0):
$0-0 < -3$
$0 < -3$ False

x	y
0	5
$\frac{5}{2}$	0

Check (0,0):
$2(0)+0 \leq 5$
$0 \leq 5$ True

This system has no solution.
There is no region where the three shaded areas overlap.

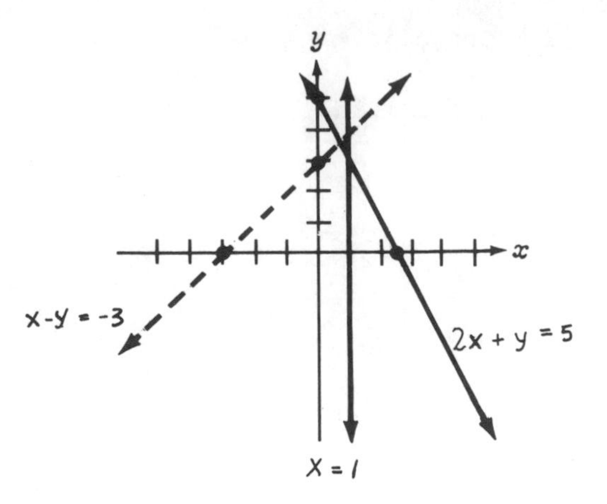

6·7 Variation

1. If $y = kx$, where k is the constant of variation, then y varies directly as x, or y is directly proportional to x.

2. If $y = k/x$, where k is the constant of variation, then y varies inversely as x, or y is inversely proportional to x.

3. If $y = kxw$, where k is the constant of variation, then y varies jointly as x and w.

4. [5] The area of a trapezoid varies jointly as its height and the sum of its bases. The area is 18 square centimeters when the height is 3 cm and the bases are 5 cm and 7 cm. Write an equation and determine the constant of variation.

$A = K(h)(b_1 + b_2)$

$18 = K(3)(5+7)$

$18 = 36K$

$K = \frac{1}{2}$

5. $x - y < -3$ and $2x + y \leq 5$ and $x \geq 1$

6·7 Variation

1. If $y = kx$, where k is the constant of variation, then y varies

   ______________ as x, or y is ___________________________ to x.

2. If $y = k/x$, where k is the constant of variation, then y varies

   ______________ as x, or y is _____________________________ to x.

3. If $y = kxw$, where k is the constant of variation, then y varies

   ______________ as x and w.

4. [5] The area of a trapezoid varies jointly as its height and the sum of its bases. The area is 18 square centimeters when the height is 3 cm and the bases are 5 cm and 7 cm. Write an equation and determine the constant of variation.

5. | 9

When an object is dropped, the distance it falls in t seconds varies directly as the square of t. If an object falls 4 feet in 1/2 second, how far will it fall in 3 seconds?

$d = Kt^2$

$4 = K(\frac{1}{2})^2$

$K = 16$

$d = Kt^2$

$d = 16(3)^2$

$\boxed{d = 144 \text{ ft.}}$

6. | 17

The volume of a certain gas varies directly as the temperature and inversely as the pressure. When the temperature is 400 K and the pressure is 12 lb/sq in., the volume is 80 cu in.. Find the pressure on the same gas when the volume is 54 cu in. and the temperature is 450 K.

$V = \frac{Kt}{P}$

$80 = \frac{K \cdot 400}{12}$

$K = 2.4$

$V = \frac{Kt}{P}$

$54 = \frac{K(450)}{P}$

$54 = \frac{2.4(450)}{P}$

$\boxed{P = 20 \text{ pounds per sq. inch}}$

5. | 9

When an object is dropped, the distance it falls in t seconds varies directly as the square of t. If an object falls 4 feet in 1/2 second, how far will it fall in 3 seconds?

6. | 17

The volume of a certain gas varies directly as the temperature and inversely as the pressure. When the temperature is 400 K and the pressure is 12 lb/sq in., the volume is 80 cu in.. Find the pressure on the same gas when the volume is 54 cu in. and the temperature is 450 K.

6.8 Branch Functions

Sketch the graphs of the following functions.

1. $f(x) = \begin{cases} 2 & \text{if } x \leq -3 \\ 4 & \text{if } -3 < x \leq 2 \\ 6 & \text{if } x > 2 \end{cases}$

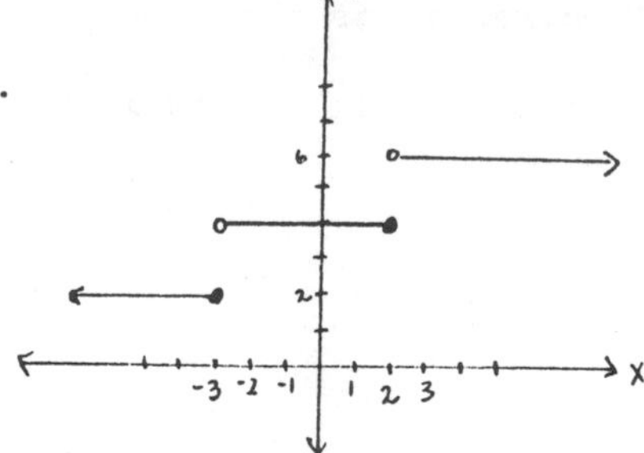

2.
11

$f(x) = \begin{cases} -3x + 1 & \text{if } x \neq 2 \\ 4 & \text{if } x = 2 \end{cases}$

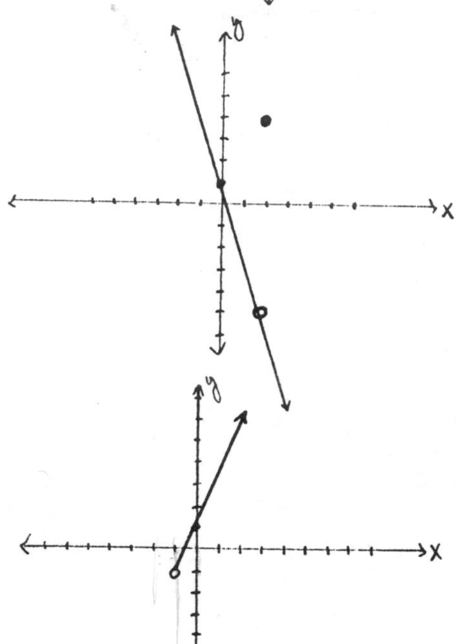

3. $f(x) = \begin{cases} 2x - 5 & \text{if } x \leq -1 \\ 2x + 1 & \text{if } x > -1 \end{cases}$

4.
25

The rates charged by Leroy's Laundromat are determined by weight. If your load is under 6 pounds Leroy will charge you $8. If your load is at least 6 pounds but less than 10 pounds the charge is $20. If the load is at least 10 pounds but no more than 14 pounds the charge is $30. (Leroy doesn't handle loads greater than 14 pounds.) Convert Leroy's charges into a branch function. Sketch the function.

$f(x) = \begin{cases} 8 & \text{if } x < 6 \\ 20 & \text{if } 6 \leq x < 10 \\ 30 & \text{if } 10 \leq x \leq 14 \end{cases}$

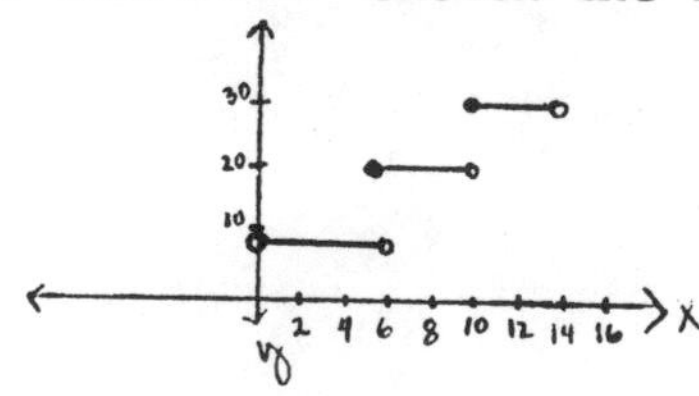

6.8 Branch Functions

Sketch the graphs of the following functions.

1. $f(x) = \begin{cases} 2 & \text{if } x \leq -3 \\ 4 & \text{if } -3 < x \leq 2 \\ 6 & \text{if } x > 2 \end{cases}$

2. | 11 $f(x) = \begin{cases} -3x + 1 & \text{if } x \neq 2 \\ 4 & \text{if } x = 2 \end{cases}$

3. $f(x) = \begin{cases} 2x - 5 & \text{if } x \leq -1 \\ 2x + 1 & \text{if } x > -1 \end{cases}$

4. | 25 The rates charged by Leroy's Laundromat are determined by weight. If your load is under 6 pounds Leroy will charge you \$8. If your load is at least 6 pounds but less than 10 pounds the charge is \$20. If the load is at least 10 pounds but no more than 14 pounds the charge is \$30. (Leroy doesn't handle loads greater than 14 pounds.) Convert Leroy's charges into a branch function. Sketch the function.

Chapter 6 Self-Test

Sketch the graphs of the following functions.

1. $y = x^2 - 5x + 6$

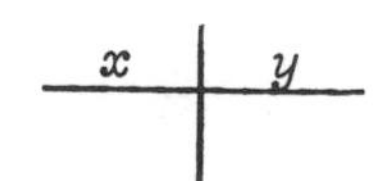

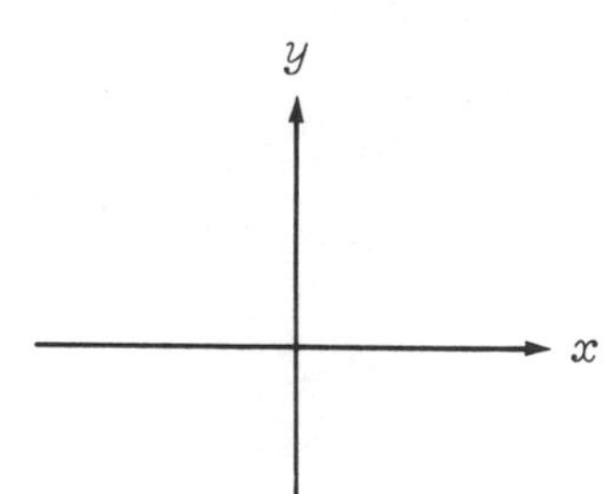

2. $y = \begin{cases} -x + 5 & \text{if } x < .5 \\ x - 1 & \text{if } x \geq .5 \end{cases}$

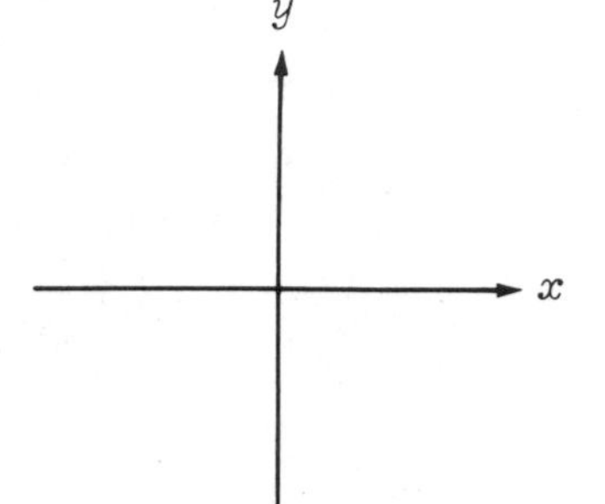

3. $3x > -2y + 6$

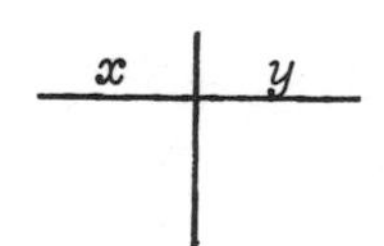

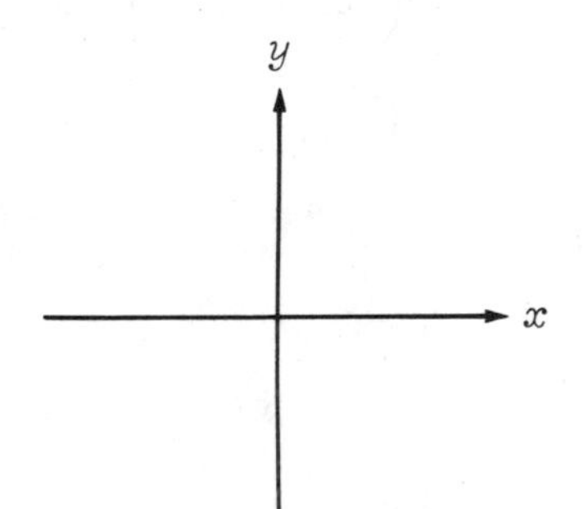

4. $3x \geq y$ and $2x - 5y < 10$

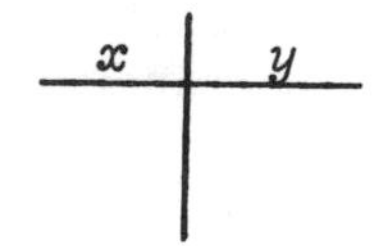

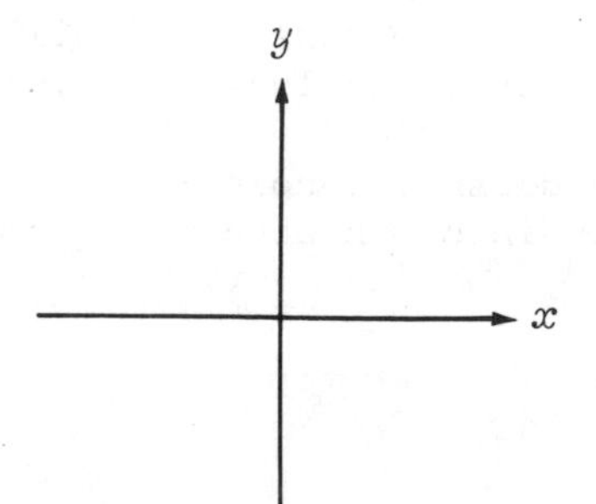

Given $f(x) = 3x^2 - 4x$ and $g(x) = 5 - 2x$, find:

5. $f(x + 1) - g(x - 1)$

6. $f[g(2)]$

7. Find the equation of the line passing through the point $(-3, 2)$ with slope $m = 3/4$.

8. Find the equation of the line passing through the points $(-5, 1/2)$ and $(3, -1)$.

9. Find the equation of the line passing through the point $(2, -6)$ and perpendicular to the line $3x - y = 4$.

10. Determine the length of the line segment between the points $(2, -5)$ and $(-1, 0)$.

11. When an object is dropped, the distance it falls in t seconds varies directly as the square of t. If an object falls 6 feet in 3/4 second, how far will it fall in 5 seconds?

C H A P T E R

7 Quadratic and Higher Order Equations and Inequalities

7 · 1 Graphing Quadratic Functions

1. Any equation that can be written in the form $y = ax^2 + bx + c$, where a, b, and c are real numbers and $a \neq 0$, is said to define a quadratic function. The graph of this function is a parabola.

2. Given a quadratic function in the form $y = 2x^2 - 8x + 11$,

   then $y = (2x^2 - 8x + 8) + 3$

   $y = 2(x^2 - 4x + 4) + 3$

   $y = 2(x - 2)^2 + 3$

   This example shows that a quadratic function in the form $y = ax^2 + bx + c$ can be written in the form $y = a(x - h)^2 + k$. In this example, $a = $ 2, $h = $ 2, and $k = $ 3. The coordinates of the vertex will be (h,k); therefore, the vertex is (2, 3).

CHAPTER

7 Quadratic and Higher Order Equations and Inequalities

7·1 Graphing Quadratic Functions

1. Any equation that can be written in the form $y = ax^2 + bx + c$, where a, b, and c are real numbers and $a \neq 0$, is said to define a ______________ ______________. The graph of this function is a ______________.

2. Given a quadratic function in the form $y = 2x^2 - 8x + 11$,

$$\text{then } y = (2x^2 - 8x + 8) + 3$$
$$y = 2(x^2 - 4x + 4) + 3$$
$$y = 2(x - 2)^2 + 3$$

This example shows that a quadratic function in the form $y = ax^2 + bx + c$ can be written in the form $y = a(x - h)^2 + k$. In this example, a = 2, h = 2, and k = 3. The coordinates of the vertex will be (h,k); therefore, the vertex is (2, 3).

3. Given a quadratic function in the form

$$y = a(x - h)^2 + k, \ a \neq 0$$

(a) The graph will be a parabola.

(b) If $a > 0$, the graph will open upward and the vertex will be a minimum or the lowest point of the function.

(c) If $a < 0$, the graph will open downward and the vertex will be a maximum or the highest point of the function.

(d) The vertical line passing through the vertex is called the axis and its equation is $x = h$.

Determine the vertex and sketch the graph of each function. Remember to rewrite each equation in the form $y = a(x - h)^2 + k$.

4. $y = 2x^2 - 5$

$y = 2(x-0)^2 - 5$

vertex: $(h,k) \rightarrow (0,-5)$

axis: $x = h \rightarrow x = 0$

Since $a > 0$, opens upward

$a = 2$, $h = 0$, $k = -5$

x	y
0	-5
±1	-3
±2	3

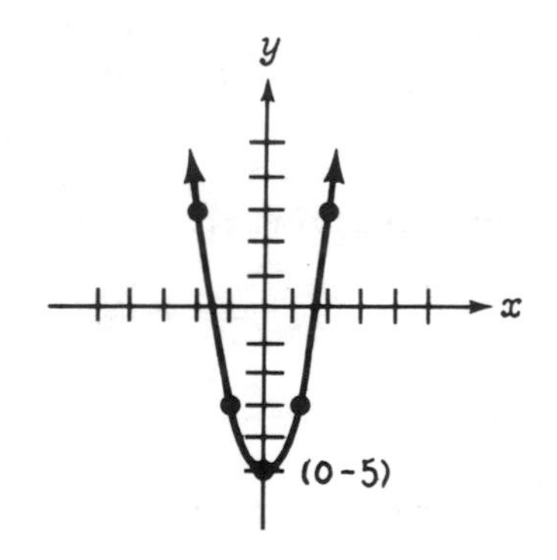

5. $y - 4 = -4x^2$

$y = -4x^2 + 4$

$y = -4(x-0)^2 + 4$

vertex: $(0,4)$

axis: $x = 0$

Since $a < 0$, opens downward

$a = -4$, $h = 0$, $k = 4$

x	y
0	4
±1	0

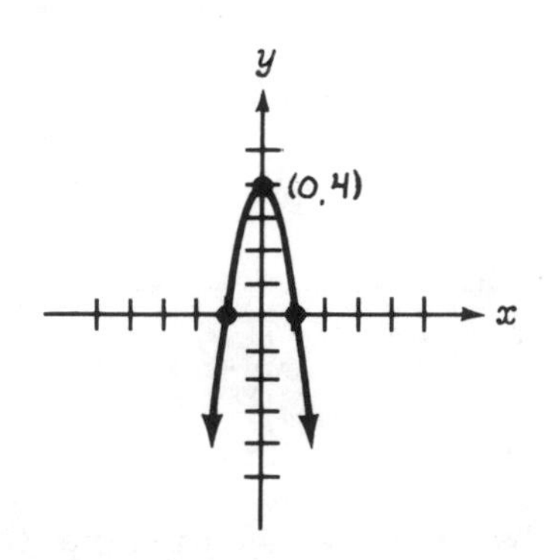

3. Given a quadratic function in the form

$$y = a(x - h)^2 + k, \; a \neq 0$$

(a) The graph will be a ____________.

(b) If $a > 0$, the graph will open __________ and the __________ will be a ___________ or the lowest point of the function.

(c) If $a < 0$, the graph will open ____________ and the __________ will be a ___________ or the highest point of the function.

(d) The vertical line passing through the vertex is called the _________ and its equation is _________.

Determine the vertex and sketch the graph of each function. Remember to rewrite each equation in the form $y = a(x - h)^2 + k$.

4. $y = 2x^2 - 5$

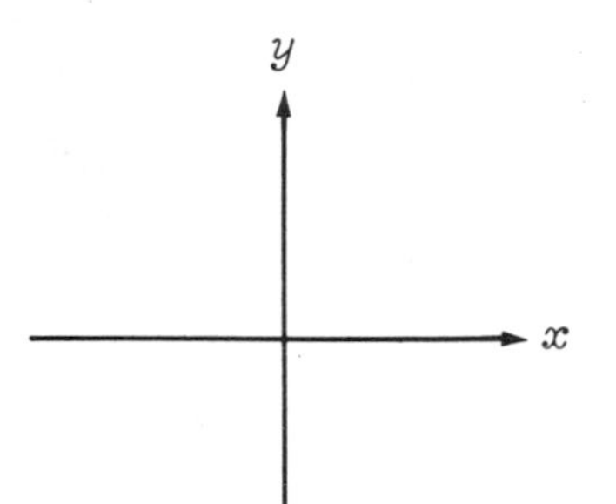

5. $y - 4 = -4x^2$

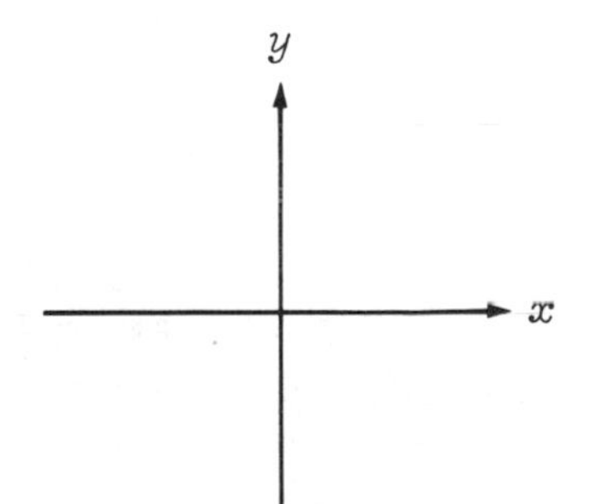

6. 21 $y = 2/3\ (x - 5)^2$ $a = \frac{2}{3}$

$y = \frac{2}{3}(x-5)^2+0$ $h=5$

vertex : (5,0) $k=0$

axis: $x=5$

Since $a>0$, opens upward

x	y
5	0
2	6

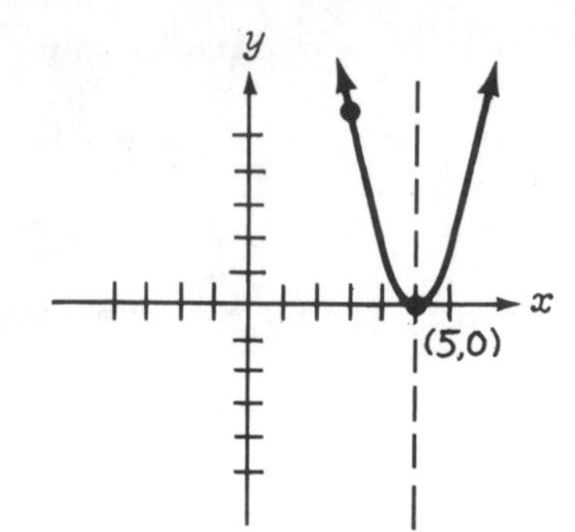

7. 35 $y = 2(x - 3)^2 - 4$ $a=2$

vertex: (3, -4) $h=3$

axis: $x=3$ $k=-4$

Since $a>0$, opens upward

x	y
2	-2
4	-2
3	-4

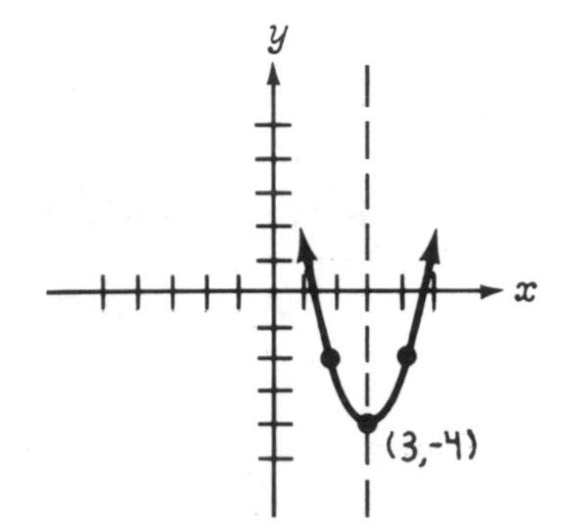

7·2 Completing the Square to Find the Vertex

1. Given: $y = x^2 + 10x + 8$

$y = (x^2 + 10x \qquad) + 8$

$y = (x^2 + 10x + \underline{25}) + 8 - \underline{25}$ (1) $(1/2 \cdot b)^2 = (1/2 \cdot 10)^2$ $= 25$

$y = (x + 5)^2 - 17$ (2) Factor the perfect-square trinomial.

The vertex is at $\underline{(-5, -17)}$. The parabola opens $\underline{\text{upward } (a=1)}$ ($a>0$).

6. 21 $y = 2/3\ (x - 5)^2$

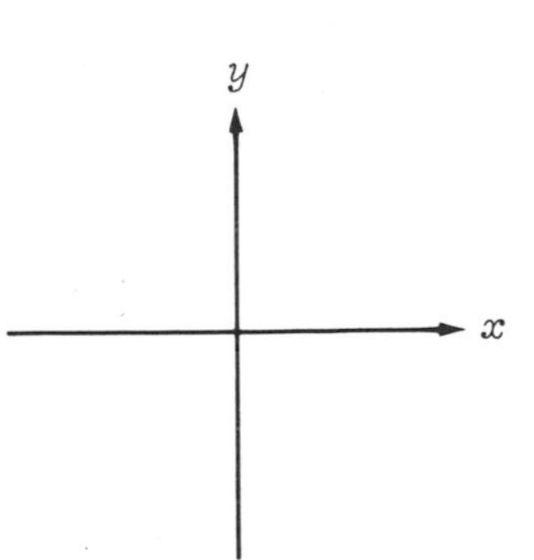

7. 35 $y = 2(x - 3)^2 - 4$

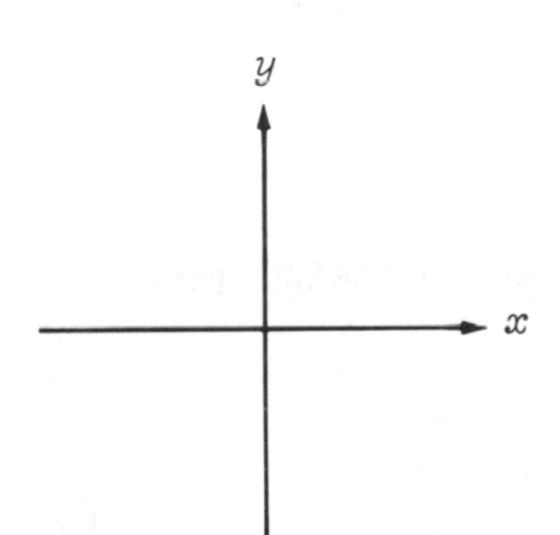

7·2 Completing the Square to Find the Vertex

1. Given: $y = x^2 + 10x + 8$

$y = (x^2 + 10x \qquad) + 8$

$y = (x^2 + 10x + \_\_\_) + 8 - \_\_\_$ (1) $(1/2 \cdot b)^2 = (1/2 \cdot 10)^2$ $= 25$

$y = (x + 5)^2 - 17$ (2) Factor the perfect-square trinomial.

The vertex is at ______________. The parabola opens ______________.

2. Given: $y = -2x^2 + 12x + 1$

$y = (-2x^2 + 12x \qquad) + 1$

$y = -2(x^2 - 6x \qquad) + 1$ (1) Factor out a, making the coefficient of x^2 equal to 1.

$y = -2(x^2 - 6x + \underline{9}) + 1 - \underline{(-18)}$ (2) $(1/2 \cdot 6)^2 = 9$ and $-2(9) = -18$

$y = -2(x - 3)^2 + 19$

The vertex is at $\underline{(3, 19)}$. The parabola opens $\underline{\text{downward } (a = -2)}$. ($a < 0$)

Put the following quadratic equations in standard form. Determine the vertex and sketch the graph.

3. $y = -x^2 + 3x + 2$

$y = -1(x^2 - 3x \qquad) + 2$

$= -1(x^2 - 3x + \frac{9}{4}) + 2 + \frac{9}{4}$

$y = -1(x - \frac{3}{2})^2 + \frac{17}{4}$

$a = -1$ vertex: $(\frac{3}{2}, \frac{17}{4})$

$h = \frac{3}{2}$ axis: $x = \frac{3}{2}$

$k = \frac{17}{4}$ opens downward, since $a < 0$

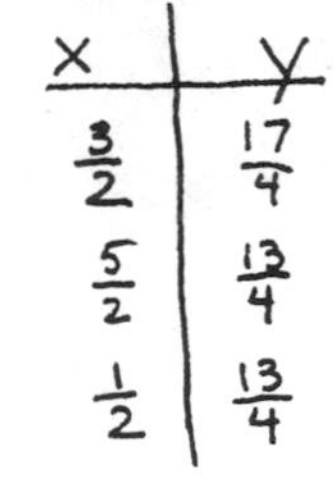

x	y
$\frac{3}{2}$	$\frac{17}{4}$
$\frac{5}{2}$	$\frac{13}{4}$
$\frac{1}{2}$	$\frac{13}{4}$

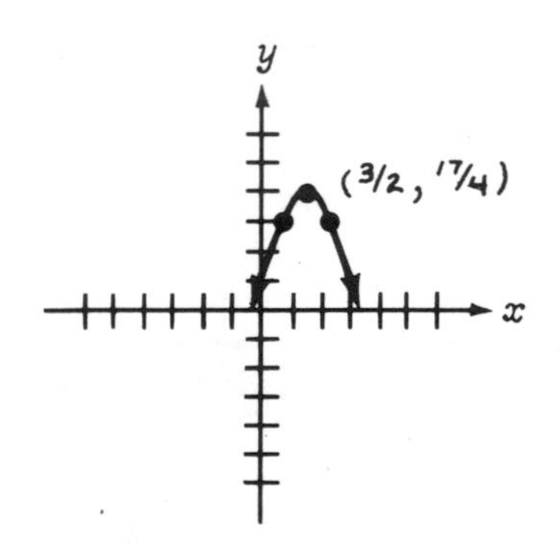

2. Given: $y = -2x^2 + 12x + 1$

$y = (-2x^2 + 12x \qquad) + 1$

$y = -2(x^2 - 6x \qquad) + 1$ (1) Factor out a, making the coefficient of x^2 equal to 1.

$y = -2(x^2 - 6x + \_\_\_) + 1 - \_\_\_$ (2) $(1/2 \cdot 6)^2 = 9$ and $-2(9) = -18$

$y = -2(x - 3)^2 + 19$

The vertex is at ____________. The parabola opens ________________.

Put the following quadratic equations in standard form. Determine the vertex and sketch the graph.

3. $y = -x^2 + 3x + 2$

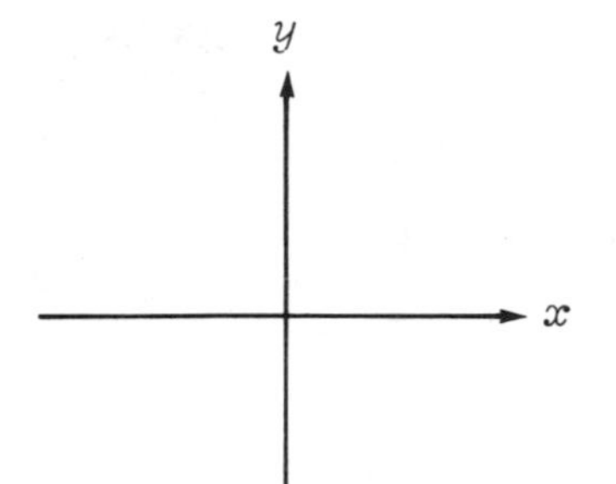

4. $y = 2x^2 - 8x + 3$

$y = 2(x^2-4x) + 3$

$\quad = 2(x^2-4x+4)+3-8$

$y = 2(x-2)^2-5$

$a=2$ vertex: $(2,-5)$

$h=2$ axis: $x=2$

$k=-5$ opens upward since $a>0$

x	y
2	-5
1	-3
3	-3

y

x

(2,-5)

5. [35] $y = -4x^2 - 40x - 92$

$y = -4(x^2+10x+25)-92+100$

$y = -4(x+5)^2+8$ $a=-4$

vertex: $(-5,8)$ $h=-5$

axis: $x=-5$ $k=8$

opens downward, since $a<0$

x	y
-5	8
-4	4
-6	4

y

(-5,8)

x

6. $y = 1/4\,x^2 + 1x + 5$

$y = \frac{1}{4}(x^2+4x \quad)+5$

$y = \frac{1}{4}(x^2+4x+4)+5-1$

$y = \frac{1}{4}(x+2)^2+4$ $a=\frac{1}{4}$

vertex: $(-2,4)$ $h=-2$

$k=4$

x	y
-2	4
0	5
-4	5

since $a>0$, opens upward

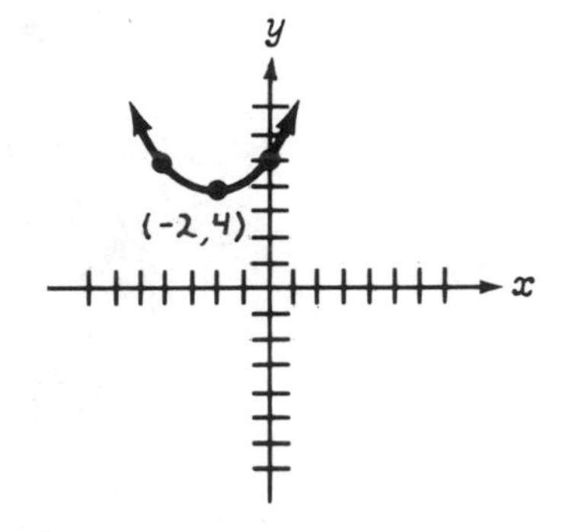

4. $y = 2x^2 - 8x + 3$

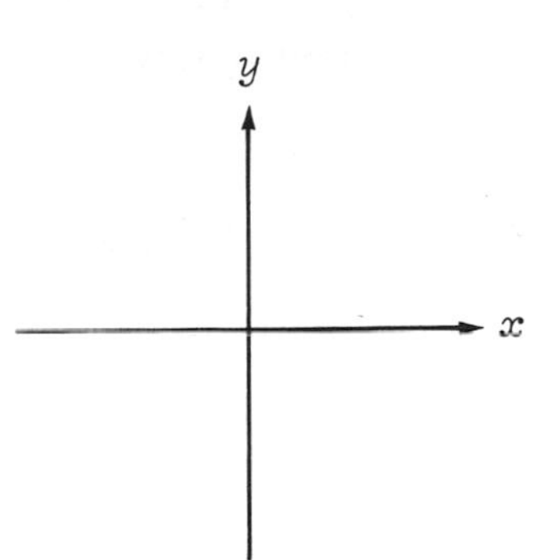

5. 35 $y = -4x^2 - 40x - 92$

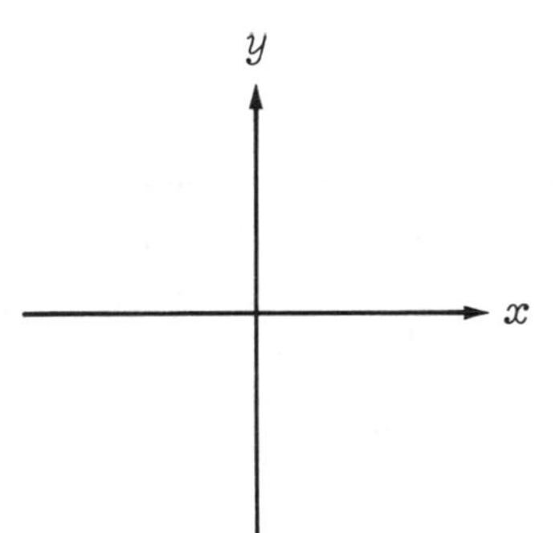

6. $y = 1/4\ x^2 + 1x + 5$

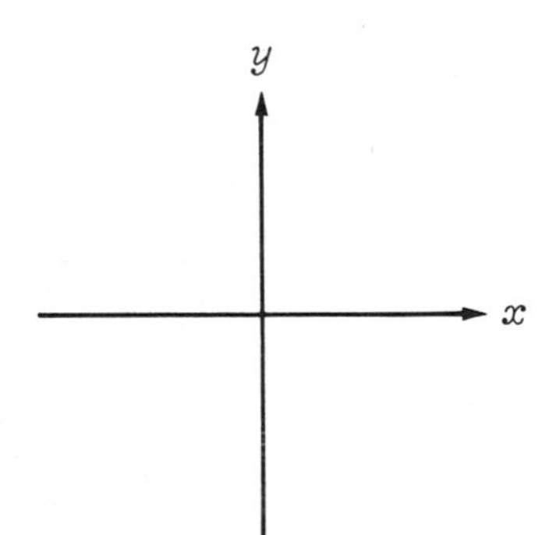

7. 57 Mitch wants to plant a rectangular garden next to a straight river. He has 120 feet of fencing and will not fence the side of the garden by the river. What dimensions will produce the largest garden?

x = width of garden
$120 - 2x$ = length of garden

Area = length × width
$= x(120-2x) = 120 - 2x^2$
$= -2(x^2 - 60)$
$= -2(x^2 - 60 + 900) + 1800$
$= -2(x-30)^2 + 1800$
vertex: $(30, 1800)$

Since $a < 0$, the graph opens downward: the vertex is a maximum, which is required in the problem.
Therefore, $x = 30$ ft = width & $120 - 2x = 60$ ft = length

7·3 Solving Quadratic Equations by Completing the Square

1. $3x^2 + 6x - 4 = 0$

$3x^2 + 6x = \underline{4}$

$\underline{3}(x^2 + 2x \quad) = 4$

$(x^2 + 2x \quad) = 4/3$

$(x^2 + 2x + \underline{1}) = 4/3 + \underline{1}$ (1) Complete the square.

$\sqrt{(\underline{x+1})^2} = \sqrt{\underline{7/3}}$ (2) Extract the roots.

$x + 1 = \pm \frac{\sqrt{7}}{\sqrt{3}} \cdot \frac{\sqrt{3}}{\sqrt{3}}$ (3) Rationalize the denominator.

$x + 1 = \pm \frac{\sqrt{21}}{3}$

$x = -1 \pm \frac{\sqrt{21}}{3}$

$x = -1 + \frac{\sqrt{21}}{3} \doteq .528$ or $x = -1 - \frac{\sqrt{21}}{3} \doteq -2.528$

7. 57 Mitch wants to plant a rectangular garden next to a straight river. He has 120 feet of fencing and will not fence the side of the garden by the river. What dimensions will produce the largest garden?

7·3 Solving Quadratic Equations by Completing the Square

1. $3x^2 + 6x - 4 = 0$

$3x^2 + 6x = \underline{4}$

$\underline{3}(x^2 + 2x + 4) = 4$

$3(x^2 + 2x + \quad) = 4/3$

$(x^2 + 2x + \underline{1}) = 4/3 + \underline{1}$ (1) Complete the square.

$\sqrt{(\underline{x+1})^2} = \sqrt{\underline{7/3}}$ (2) Extract the roots.

$x + 1 = \pm \underline{\frac{\sqrt{7}}{\sqrt{3}} \cdot \frac{\sqrt{3}}{\sqrt{3}}}$ (3) Rationalize the denominator.

$x + 1 = \pm \underline{\frac{\sqrt{21}}{3}}$

$x = \underline{-1 \pm \frac{\sqrt{21}}{3}}$

$x = \underline{-1 + \frac{\sqrt{21}}{3} \doteq .528}$ or $x = \underline{-1 - \frac{\sqrt{21}}{3} \doteq -2.528}$

Solve the following quadratic equations by extracting the roots.

2. $x^2 = 300$

$\sqrt{x^2} = \sqrt{300}$

$x = \sqrt{100} \cdot \sqrt{3}$

$\boxed{x = \pm 10\sqrt{3}}$

$x \doteq \pm 17.3$

3. $3x^2 = 5$

$x^2 = \frac{5}{3}$

$\sqrt{x^2} = \sqrt{\frac{5}{3}}$

$\boxed{x = \frac{\sqrt{5}}{\sqrt{3}} \cdot \frac{\sqrt{3}}{\sqrt{3}} = \frac{\sqrt{15}}{3}}$

$x \doteq \pm 1.29$

4. / 25 $9x^2 + 1 = 5$

$9x^2 = 4$

$x^2 = \frac{4}{9}$

$\sqrt{x^2} = \sqrt{\frac{4}{9}}$

$\boxed{x = \pm \frac{2}{3}}$

5. / 41 $(x - 1)^2 = -75$

$\sqrt{(x-1)^2} = \sqrt{-75}$

$x - 1 = \pm i\sqrt{75}$

$= \pm i\sqrt{25} \cdot \sqrt{3}$

$= \pm 5i\sqrt{3}$

$x - 1 = \pm 5i\sqrt{3}$

$\boxed{x = 1 \pm 5i\sqrt{3}}$

Find the solutions of the following quadratic equations by completing the square.

6. $x^2 + 9x + 20 = 0$

$x^2 + 9x \quad = -20$

$x^2 + 9x + \frac{81}{4} = -20 + \frac{81}{4}$

$(x + \frac{9}{2})^2 = \frac{1}{4}$

$\sqrt{(x + \frac{9}{2})^2} = \sqrt{\frac{1}{4}}$

$x + \frac{9}{2} = \pm \frac{1}{2}$

$x = -\frac{9}{2} \pm \frac{1}{2}$

$\boxed{x = -4 \text{ OR } x = -5}$

7. $3x^2 + 27x = 0$

$3(x^2 + 9x \quad) = 0$

$x^2 + 9x + \frac{81}{4} = \frac{81}{4}$

$\sqrt{(x + \frac{9}{2})^2} = \sqrt{\frac{81}{4}}$

$x + \frac{9}{2} = \pm \frac{9}{2}$

$x = -\frac{9}{2} \pm \frac{9}{2}$

$\boxed{x = 0 \quad \text{OR} \quad x = -9}$

Solve the following quadratic equations by extracting the roots.

2. $x^2 = 300$

$\sqrt{x^2} = \sqrt{300}$

$x = \sqrt{100} \cdot \sqrt{3}$

$x = \pm 10\sqrt{3}$

$x = \pm 17.3$

3. $3x^2 = 5$

$\sqrt{x^2} = \sqrt{\frac{5}{3}}$

$x = \frac{\sqrt{5}}{\sqrt{3}} \cdot \frac{\sqrt{3}}{\sqrt{3}} = \frac{\sqrt{15}}{3}$

$x = \frac{\sqrt{15}}{3}$

$x \doteq \pm 1.29$

$\frac{3.873}{3}$

4. (25) $9x^2 + 1 = 5$

$9x^2 = 5 - 1$

$\frac{9x^2}{9} = \frac{4}{9}$

$\sqrt{x^2} = \sqrt{\frac{4}{9}} \cdot \frac{\sqrt{4}}{\sqrt{4}} = \frac{\sqrt{16}}{\sqrt{36}}$

$x = \frac{4}{6}$

$x = \pm\frac{2}{3}$

5. (41) $(x - 1)^2 = -75$

$\sqrt{(x-1)^2} = \sqrt{-75}$

$x - 1 = i\sqrt{25 \cdot 3}$

$x - 1 = i\,5\sqrt{3}$

$x = 1 + i5\sqrt{3}$

or $x = 1 - i5\sqrt{3}$

Find the solutions of the following quadratic equations by completing the square.

6. $x^2 + 9x + 20 = 0$

$x^2 + 9x + \frac{81}{4} = -20 + \frac{81}{4}$

$(x + \frac{9}{2})^2 = \frac{-80}{4} + \frac{81}{4}$

$\sqrt{(x + \frac{9}{2})^2} = \sqrt{\frac{1}{4}}$

$x + \frac{9}{2} = \pm\frac{1}{2}$

$x = \frac{9}{2} + \frac{1}{2}$ $\quad x = \frac{9}{2} - \frac{1}{2}$

$x = \frac{10}{2}$ $\quad x = \frac{8}{2}$

$x = 5$ $\quad x = 4$

7. $3x^2 + 27x = 0$

$3(x^2 + 9x) = 0$

$x + 9x + \frac{81}{4} = \frac{0}{3} + \frac{81}{4}$

$\sqrt{(x + \frac{9}{2})^2} = \sqrt{\frac{81}{4}}$

$x + \frac{9}{2} = \pm\frac{9}{2}$

$x = \frac{9}{2} + \frac{9}{2}$ $\quad x = \frac{9}{2} - \frac{9}{2}$

$x = \frac{18}{2}$ $\quad x = 0$

$x = -9$

why negative

8. (87) $16x^2 - 8x - 7 = 0$

$x^2 - \frac{1}{2}x - \frac{7}{16} = 0$

$(x^2 - \frac{1}{2}x \qquad) = \frac{7}{16}$

$x^2 - \frac{1}{2}x + \frac{1}{16} = \frac{7}{16} + \frac{1}{16}$

$\sqrt{(x - \frac{1}{4})^2} = \sqrt{\frac{8}{16}}$

$x - \frac{1}{4} = \frac{\pm\sqrt{8}}{4} = \frac{\pm 2\sqrt{2}}{4}$

$\boxed{x = \frac{1 \pm 2\sqrt{2}}{4}}$

9. $x^2 - 4x + 6 = 0$

$x^2 - 4x = -6$

$x^2 - 4x + 4 = -6 + 4$

$\sqrt{(x-2)^2} = \sqrt{-2}$

$x - 2 = \pm i\sqrt{2}$

$\boxed{x = 2 \pm i\sqrt{2}}$

10. $3x^2 + 12x = -7$

$x^2 + 4x = -\frac{7}{3}$

$x^2 + 4x + 4 = -\frac{7}{3} + 4$

$\sqrt{(x+2)^2} = \sqrt{\frac{5}{3}}$

$x + 2 = \pm \frac{\sqrt{15}}{3}$

$\boxed{x = -2 \pm \frac{\sqrt{15}}{3}}$

11. (97) $36x^2 + 96x + 73 = 0$

$x^2 + \frac{96}{36}x = -\frac{73}{36}$

$x^2 + \frac{96}{36}x + \frac{2304}{1296} = \frac{-73}{36} + \frac{2304}{1296}$

$(x + \frac{48}{36})^2 = \frac{-9}{36}$

$\sqrt{(x + \frac{4}{3})^2} = \sqrt{\frac{-9}{36}}$

$x + \frac{4}{3} = \pm \frac{1}{2}i$

$\boxed{x = -\frac{4}{3} \pm \frac{1}{2}i}$

7.4 The Quadratic Formula

1. If $ax^2 + bx + c = 0$, where a, b, and c are real numbers and $a \neq 0$, then the solutions of the equation can be found by use of the <u>quadratic</u> <u>formula</u> which is: $x = \frac{-b \pm \sqrt{b^2 - 4ac}}{2a}$.

8. (87) $16x^2 - 8x - 7 = 0$

9. $x^2 - 4x + 6 = 0$

10. $3x^2 + 12x = -7$

11. (97) $36x^2 + 96x + 73 = 0$

7·4 The Quadratic Formula

1. If $ax^2 + bx + c = 0$, where a, b, and c are real numbers and $a \neq 0$, then the solutions of the equation can be found by use of the

_______________ _______________ which is: x = _____________________.

2. Given: $2x^2 - 3x = -5$

$2x^2 - 3x + 5 = 0$ (1) Write in standard form.

$x = \dfrac{-b \pm \sqrt{b^2 - 4ac}}{2a}$ (2) Substitute values into formula and simplify.

$a = 2$ $\quad x = \dfrac{-(\underline{-3}) \pm \sqrt{(\underline{-3})^2 - 4(\underline{2})(\underline{5})}}{2(\underline{2})}$

$b = -3$ $\quad x = \dfrac{3 \pm \sqrt{\underline{9-40}}}{4}$

$c = 5$

$x = \dfrac{3 \pm \underline{i}\sqrt{\underline{31}}}{4}$

Therefore, $x = \underline{\dfrac{3+i\sqrt{31}}{4}}$ or $x = \underline{\dfrac{3-i\sqrt{31}}{4}}$

3. The expression $b^2 - 4ac$ is called the discriminant. Notice that this is the value found under the radical sign in the quadratic formula.

(a) If $b^2 - 4ac > 0$ and is a perfect square, then there are 2 rational solutions of the equation; if $b^2 - 4ac$ is not a perfect square, there are 2 irrational solutions.

(b) If $b^2 - 4ac = 0$, there is 1 rational solution.

(c) If $b^2 - 4ac < 0$, there are 2 imaginary solutions.

Use the discriminant to characterize the solutions of the following quadratic equations.

4. $x^2 - 3x + 5 = 0$

$a = \underline{1}$ $b = \underline{-3}$ $c = \underline{5}$

Discriminant value: $(-3)^2 - 4(1)(5) = -11$

Number of solutions: 2

Nature of solutions: imaginary

5. $2x = -7x^2 + 1$

$a = \underline{7}$ $b = \underline{2}$ $c = \underline{-1}$

$7x^2 + 2x - 1 = 0$

Discriminant value: $b^2 - 4ac = 2^2 - 4(7)(-1) = 32$

Number of solutions: 2

Nature of solutions: irrational

2. Given: $2x^2 - 3x = -5$

$2x^2 - 3x + 5 = 0$ (1) Write in standard form.

$x = \frac{-b \pm \sqrt{b^2 - 4ac}}{2a}$ (2) Substitute values into formula and simplify.

$a = 2$ $x = \frac{-(\_\_) \pm \sqrt{(\_\_)^2 - 4(\_\_)(\_\_)}}{2(\_\_)}$

$b = -3$

$c = 5$ $x = \frac{3 \pm \sqrt{\_\_\_\_}}{4}$

$x = \frac{3 \pm \_\_\sqrt{\_\_\_\_}}{4}$

Therefore, $x =$ ______________ or $x =$ ______________

3. The expression $b^2 - 4ac$ is called the ________________. Notice that this is the value found under the radical sign in the quadratic formula.

   (a) If $b^2 - 4ac > 0$ and is a perfect square, then there are ___ ____________ solutions of the equation; if $b^2 - 4ac$ is not a perfect square, there are _____ ________________ solutions.

   (b) If $b^2 - 4ac = 0$, there is _____ ________________ solution.

   (c) If $b^2 - 4ac < 0$, there are _____ ________________ solutions.

Use the discriminant to characterize the solutions of the following quadratic equations.

4. $x^2 - 3x + 5 = 0$ Discriminant value: ________________

   $a =$ ___ $b =$ ___ $c =$ ___ Number of solutions: ________________

   Nature of solutions: ________________

5. $2x = -7x^2 + 1$ Discriminant value: ________________

   $a =$ ___ $b =$ ___ $c =$ ___ Number of solutions: ________________

   Nature of solutions: ________________

6\. $x^2 - 9 = 0$ [5]

$a =$ 1 $b =$ 0 $c =$ -9

Discriminant value: $0^2-4(1)(-9)=36$

Number of solutions: 2

Nature of solutions: rational

7\. $6x - 9x^2 = 1$

$a =$ -9 $b =$ 6 $c =$ -1

$-9x^2+6x-1=0$

Discriminant value: $6^2-4(-9)(-1)=0$

Number of solutions: 1

Nature of solutions: rational

Find the solutions of each of the following quadratic equations by use of the quadratic formula.

8\. $x^2 - 3x + 5 = 0$

$a =$ 1 $b =$ -3 $c =$ 5

$$x = \frac{-b \pm \sqrt{b^2-4ac}}{2a}$$

$$= \frac{-(-3) \pm \sqrt{(-3)^2-4(1)(5)}}{2(1)}$$

$$= \frac{3 \pm \sqrt{9-20}}{2}$$

$$\boxed{x = \frac{3 \pm i\sqrt{11}}{2}}$$

9\. $2x = -7x^2 + 1$ $\quad 7x^2+2x-1=0$

$a =$ 7 $b =$ 2 $c =$ -1

$$x = \frac{-b \pm \sqrt{b^2-4ac}}{2a}$$

$$= \frac{-2 \pm \sqrt{2^2-4(7)(-1)}}{2(7)}$$

$$= \frac{-2 \pm \sqrt{4+28}}{14} = \frac{-2 \pm \sqrt{32}}{14} =$$

$$x = \frac{-2 \pm 4\sqrt{2}}{14} = \boxed{\frac{-1 \pm 2\sqrt{2}}{7}}$$

10\. $x^2 - 9 = 0$

$x^2+0x-9=0$

$a =$ 1 $b =$ 0 $c =$ -9

$$x = \frac{-b \pm \sqrt{b^2-4ac}}{2a}$$

$$= \frac{-0 \pm \sqrt{0^2-4(1)(-9)}}{2(1)}$$

$$= \frac{\pm\sqrt{36}}{2}$$

$$\boxed{x = \pm 3}$$

11\. $6x - 9x^2 = 1$

$-9x^2+6x-1=0$

$a =$ -9 $b =$ 6 $c =$ -1

$$x = \frac{-b \pm \sqrt{b^2-4ac}}{2a}$$

$$= \frac{-6 \pm \sqrt{(6)^2-4(-9)(-1)}}{2(-9)}$$

$$= \frac{-6 \pm \sqrt{36-36}}{-18}$$

$$\boxed{x = \frac{1}{3}}$$

6. | 5 $x^2 - 9 = 0$

$a =$ ___ $b =$ ___ $c =$ ___

Discriminant value: ______________

Number of solutions: ______________

Nature of solutions: ______________

7. $6x - 9x^2 = 1$

$a =$ ___ $b =$ ___ $c =$ ___

Discriminant value: ______________

Number of solutions: ______________

Nature of solutions: ______________

Find the solutions of each of the following quadratic equations by use of the quadratic formula.

8. $x^2 - 3x + 5 = 0$

$a =$ ___ $b =$ ___ $c =$ ___

9. $2x = -7x^2 + 1$

$a =$ ___ $b =$ ___ $c =$ ___

10. $x^2 - 9 = 0$

$a =$ ___ $b =$ ___ $c =$ ___

11. $6x - 9x^2 = 1$

$a =$ ___ $b =$ ___ $c =$ ___

12. 27

$x(3x - 1) = -1$

$3x^2 - x + 1 = 0$

$a = \underline{3}$ $b = \underline{-1}$ $c = \underline{1}$

$$x = \frac{-b \pm \sqrt{b^2 - 4ac}}{2a}$$

$$= \frac{1 \pm \sqrt{(-1)^2 - 4(3)(1)}}{2(3)}$$

$$\boxed{x = \frac{1 \pm i\sqrt{11}}{6}}$$

13. 37

$1/2 - 1/x - 1/x^2 = 0$

NOTE: Multiply both sides by $2x^2$

$x^2 - 2x - 2 = 0$

$a = \underline{1}$ $b = \underline{-2}$ $c = \underline{-2}$

$$x = \frac{-b \pm \sqrt{b^2 - 4ac}}{2a}$$

$$= \frac{2 \pm \sqrt{(-2)^2 - 4(1)(-2)}}{2(1)}$$

$$x = \frac{2 \pm \sqrt{4+8}}{2} = \frac{2 \pm 2\sqrt{3}}{2} = \boxed{1 \pm \sqrt{3}}$$

7·5 Applications

1. 11 The base of a triangle is one inch less than twice the height. Find the base and the height if the area of the triangle is 6 square inches. (Hint: The area of a triangle = 1/2 • base • height.)

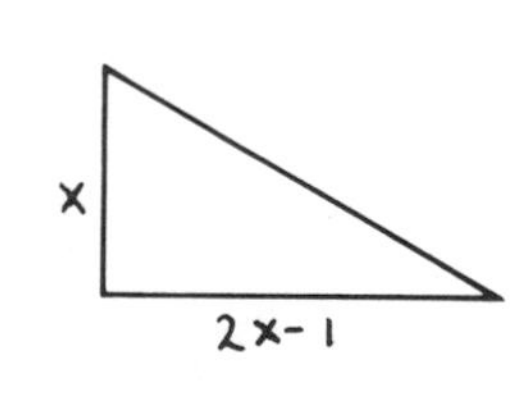

$A = \frac{1}{2} b \cdot h$

$6 = \frac{1}{2}(x)(2x-1)$

$6 = x^2 - \frac{1}{2}x$

$12 = 2x^2 - x$

$2x^2 - x - 12 = 0$

$a = 2,\ b = -1,\ c = -12$

$$x = \frac{-b \pm \sqrt{b^2 - 4ac}}{2a}$$

$$= \frac{1 \pm \sqrt{(-1)^2 - 4(2)(-12)}}{2(2)}$$

$$x = \frac{1 \pm \sqrt{1+96}}{4} = \frac{1 + \sqrt{97}}{4}$$

$x \doteq 2.71$ inches height

$2x - 1 \doteq 4.42$ inches base

NOTE: $x = \frac{1 - \sqrt{97}}{4}$ is extraneous because it yields a negative result.

2. 3 The sum of a number and its reciprocal is 4. What is the number?

x = number

$x + \frac{1}{x} = 4$

$x^2 + 1 = 4x$

$x^2 - 4x + 1 = 0$

$a = 1,\ b = -4,\ c = 1$

$$x = \frac{-b \pm \sqrt{b^2 - 4ac}}{2a} = \frac{4 \pm \sqrt{16-4}}{2}$$

$$x = \frac{4 \pm \sqrt{12}}{2} = \frac{4 \pm 2\sqrt{3}}{2}$$

$$\boxed{x = 2 + \sqrt{3} \text{ OR } 2 - \sqrt{3}}$$

12. | 27

$x(3x - 1) = -1$

a = ___ b = ___ c = ___

13. | 37

$1/2 - 1/x - 1/x^2 = 0$

a = ___ b = ___ c = ___

7·5 Applications

1. | 11

The base of a triangle is one inch less than twice the height. Find the base and the height if the area of the triangle is 6 square inches. (Hint: The area of a triangle = 1/2 • base • height.)

2. | 3

The sum of a number and its reciprocal is 4. What is the number?

3. **7** The square of a number minus two times the number is 14. What is the number?

x = the number

$x^2 - 2x = 14$

$x^2 - 2x - 14 = 0$

$a = 1, b = -2, c = -14$

$x = \frac{-b \pm \sqrt{b^2 - 4ac}}{2a}$

$x = \frac{2 \pm \sqrt{4+56}}{2} = \frac{2 \pm 2\sqrt{15}}{2}$

$\boxed{x = 1 + \sqrt{15} \text{ OR } 1 - \sqrt{15}}$

4. **19** One pipe fills a reservoir 1 hour quicker than another pipe. Together, they fill the reservoir in 4 hours. How long does it take each pipe to fill the reservoir?

x = ONE pipe's time

$x+1$ = OTHER pipe's time

$\frac{1}{x} + \frac{1}{x+1} = \frac{1}{4}$

$4(x+1) + 4x = x(x+1)$

$4x + 4 + 4x = x^2 + x$

$x^2 - 7x - 4 = 0$

$a = 1, b = -7, c = -4$

$x = \frac{-b \pm \sqrt{b^2 - 4ac}}{2a}$

$x = \frac{7 \pm \sqrt{49+16}}{2} = \frac{7 \pm \sqrt{65}}{2}$

NOTE: $x = \frac{7 - \sqrt{65}}{2}$ is extraneous.

$x = \frac{7 + \sqrt{65}}{2} \doteq 7.53$ hours

$x + 1 \doteq 8.53$ hours

5. **23** Chuck hikes up a 4-mile mountain trail. His speed going down the trail is 1 mph faster than his speed going up the trail. He spends a total of 5 hours hiking. What is Chuck's hiking speed up the trail and what is his hiking speed down the trail?

	t	r	d
up	$\frac{4}{x}$	x	4
down	$\frac{4}{x+1}$	$x+1$	4

$\frac{4}{x} + \frac{4}{x+1} = 5$

$4(x+1) + 4x = 5x(x+1)$

$4x + 4 + 4x = 5x^2 + 5x$

$5x^2 - 3x - 4 = 0$

$a = 5, b = -3, c = -4$

$x = \frac{-b \pm \sqrt{b^2 - 4ac}}{2a}$

$x = \frac{3 \pm \sqrt{9+80}}{10} = \frac{3 + \sqrt{89}}{10}$

$x \doteq 1.24$ mph up

$x + 1 \doteq 2.24$ mph down

3. | 7

The square of a number minus two times the number is 14. What is the number?

4. | 19

One pipe fills a reservoir 1 hour quicker than another pipe. Together, they fill the reservoir in 4 hours. How long does it take each pipe to fill the reservoir?

5. | 23

Chuck hikes up a 4-mile mountain trail. His speed going down the trail is 1 mph faster than his speed going up the trail. He spends a total of 5 hours hiking. What is Chuck's hiking speed up the trail and what is his hiking speed down the trail?

6. / 27

A ball is thrown upward from the roof of an 80-foot tall building with a speed of 32 feet per second. The equation that gives the ball's height above ground level is $h = -16t^2 + 32t + 80$. When does the ball hit the ground? ($h=0$)

$$0 = -16t^2 + 32t + 80$$
$$0 = t^2 - 2t - 5$$
$$a = 1,\ b = -2,\ c = -5$$

$$x = \frac{-b \pm \sqrt{b^2 - 4ac}}{2a}$$

$$= \frac{2 \pm \sqrt{4+20}}{2} = \frac{2 \pm 2\sqrt{6}}{2}$$

$$x = 1 \pm \sqrt{6}$$

$x = 1 + \sqrt{6}$, $x \doteq 3.449$ sec. OR $x = 1 - \sqrt{6}$ (extraneous)

7·6 Equations in Quadratic Form and Higher Order Equations

1. Given: $x - 7\sqrt{x} - 8 = 0$

Let $\sqrt{x} = w$, then $x = w^2$.

$w^2 - 7w - 8 = 0$		(1) Substitute.
$(w-8)(w+1) = 0$		(2) Factor.
$w-8 = 0$ or	$w+1 = 0$	(3) Solve.
$w = 8$	$w = -1$	

Since $w = \sqrt{x}$,

$(\sqrt{x})^2 = (8)^2$ or	$(\sqrt{x})^2 = (-1)^2$	(4) Substitute.
$x = 64$	$x = 1$	
$64 - 7\sqrt{64} - 8 \overset{?}{=} 0$ or	$1 - 7\sqrt{1} - 8 \overset{?}{=} 0$	(5) Check.
$64 - 56 - 8 \overset{?}{=} 0$	$1 - 7 - 8 \overset{?}{=} 0$	
$0 = 0$	$-14 \neq 0$	$x=1$ is extraneous

This example illustrates the use of the method of substitution. This method is used to convert equations of higher or lower order to quadratic form.

6. 27 A ball is thrown upward from the roof of an 80-foot tall building with a speed of 32 feet per second. The equation that gives the ball's height above ground level is $h = -16t^2 + 32t + 80$. When does the ball hit the ground?

7·6 Equations in Quadratic Form and Higher Order Equations

1. Given: $x - 7\sqrt{x} - 8 = 0$

Let $\sqrt{x} = w$, then $x = w^2$.

$w^2 - 7w - 8 = 0$ (1) Substitute.

$(\_\_\_\_\_\_)(\_\_\_\_\_\_) = 0$ (2) Factor.

$\_\_\_\_\_\_ = 0$ or $\_\_\_\_\_\_ = 0$ (3) Solve.

$w = 8$ $\quad$ $w = -1$

Since $w = \sqrt{x}$,

$\_\_\_ = 8$ or $\_\_\_ = -1$ (4) Substitute.

$\_\_\_ = 64$ $\quad$ $\_\_\_ = 1$

$\_\_\_ - 7\sqrt{\_\_\_} - 8 \stackrel{?}{=} 0$ or $\_\_\_ - 7\sqrt{\_\_\_} - 8 \stackrel{?}{=} 0$ (5) Check.

$\_\_\_\_\_\_\_\_ \stackrel{?}{=} 0$ $\quad$ $\_\_\_\_\_\_\_\_ \stackrel{?}{=} 0$

$\_\_\_\_ = 0$ $\quad$ $\_\_\_\_ = 0$

This example illustrates the use of the method of substitution. This method is used to convert equations of higher or lower order to quadratic form.

Find the solutions of each of the following equations. Use the most appropriate method to solve each problem.

2. $3x^4 - 16x^2 + 5 = 0$

$(3x^2-1)(x^2-5) = 0$

$3x^2-1=0$ OR $x^2-5=0$

$3x^2=1$ $\quad$ $x^2=5$

$x^2=\frac{1}{3}$

$$x = \frac{\pm\sqrt{3}}{3} \text{ OR } x = \pm\sqrt{5}$$

3. $x + 9\sqrt{x} = -20$ $\quad$ Let $\sqrt{x}=w$

$w^2+9w+20=0$

$a=1$ $\quad$ $w = \frac{-b\pm\sqrt{b^2-4ac}}{2a}$

$b=9$

$c=20$ $\quad$ $= \frac{-9\pm\sqrt{9^2-4(1)(20)}}{2}$

$= \frac{-9\pm\sqrt{81-80}}{2} = \frac{-9\pm 1}{2}$

Neither solution checks.

$w=-4$ OR $w=-5$

$\sqrt{x}=-4$ $\quad$ $\sqrt{x}=-5$

No solution

4. (23) $4(x - 1)^2 - 3(x - 1) - 10 = 0$

Let $(x-1)=y$

$4y^2-3y-10=0$

$(4y+5)(y-2)=0$

$4y+5=0$ $\quad$ $y-2=0$

$y=-\frac{5}{4}$ $\quad$ $y=2$

$x-1=-\frac{5}{4}$ $\quad$ $x-1=2$

$$x=-\frac{1}{4} \text{ OR } x=3$$

5. (31) $x^{-2} + 6x^{-1} + 5 = 0$

Let $x^{-1}=y$

$y^2+6y+5=0$

$(y+5)(y+1)=0$

$y+5=0$ OR $y+1=0$

$y=-5$ $\quad$ $y=-1$

$(x^{-1})^{-1}=(-5)^{-1}$ $\quad$ $x^{-1}=-1$

$$x=-\frac{1}{5} \text{ OR } x=-1$$

6. $4x^3 - 8x^2 - x + 2 = 0$

$4x^2(x-2)-(x-2)=0$

$(x-2)(4x^2-1=0$

$x-2=0$ OR $4x^2-1=0$

$$x=2 \text{ OR } x=\pm\frac{1}{2}$$

7. (69) $x^4 + 3x^3 + 27x + 81 = 0$

$x^3(x+3)+27(x+3)=0$

$(x^3+27)(x+3)=0$

$(x+3)(x^2-3x+9)(x+3)=0$

$a=1,\ b=-3,\ c=9$

$x=\frac{3\pm\sqrt{9-36}}{2}$

$$x=\frac{3\pm 3i\sqrt{3}}{2} \text{ OR } x=-3$$

Find the solutions of each of the following equations. Use the most appropriate method to solve each problem.

2. $3x^4 - 16x^2 + 5 = 0$

3. $x + 9\sqrt{x} = -20$

4. | 23 $4(x - 1)^2 - 3(x - 1) - 10 = 0$

5. | 31 $x^{-2} + 6x^{-1} + 5 = 0$

6. $4x^3 - 8x^2 - x + 2 = 0$

7. | 69 $x^4 + 3x^3 + 27x + 81 = 0$

8. $\left(\frac{3x + 1}{x}\right)^2 + 6\left(\frac{3x + 1}{x}\right) + 8 = 0$ Let $y = \left(\frac{3x+1}{x}\right)$

$y^2 + 6y + 8 = 0$

$(y+2)(y+4) = 0$

$y = -2$ OR $y = -4$

$\frac{3x+1}{x} = -2$ $\frac{3x+1}{x} = -4$

$3x+1 = -2x$ $3x+1 = -4x$

$1 = -5x$ $1 = -7x$

$\boxed{x = -\frac{1}{5} \text{ OR } x = -\frac{1}{7}}$

7 · 7 Polynomial and Rational Inequalities

1. Given: $2x^2 + 9x - 5 \geq 0$

$(\underline{2x-1})(\underline{x+5}) \geq 0$ (1) Factor.

$\underline{2x-1} = 0$ or $\underline{x+5} = 0$ (2) Determine boundary points.

$\underline{x} = \underline{\frac{1}{2}}$ or $\underline{x} = \underline{-5}$

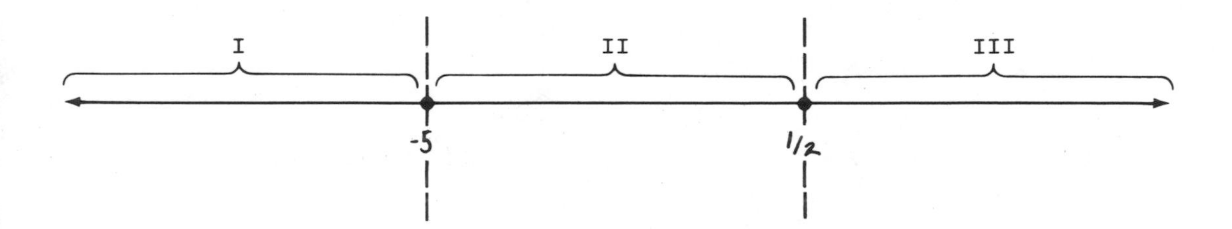

Test $x = -8$

$2(\underline{-8})^2+9(\underline{-8})-5\geq 0$

$\underline{51} \geq 0$ TRUE

Region I $\underline{\text{does}}$

contain solutions.

Test $x = 0$

$2(\underline{0})^2+9(\underline{0})-5\geq 0$

$\underline{-5} \geq 0$ FALSE

Region II $\underline{\text{does}}$ $\underline{\text{not}}$

contain solutions.

Test $x = 2$

$2(\underline{2})^2+9(\underline{2})-5\geq 0$

$\underline{21} \geq 0$ TRUE

Region III $\underline{\text{does}}$

contain solutions.

8. $\left(\frac{3x + 1}{x}\right)^2 + 6\left(\frac{3x + 1}{x}\right) + 8 = 0$

7·7 Polynomial and Rational Inequalities

1. Given: $2x^2 + 9x - 5 \geq 0$

(2x − 1)(x + 5) ≥ 0 — (1) Factor.

2(½) − 1 = 0 or −5 + 5 = 0 — (2) Determine boundary points.

x = ½ or x = −5

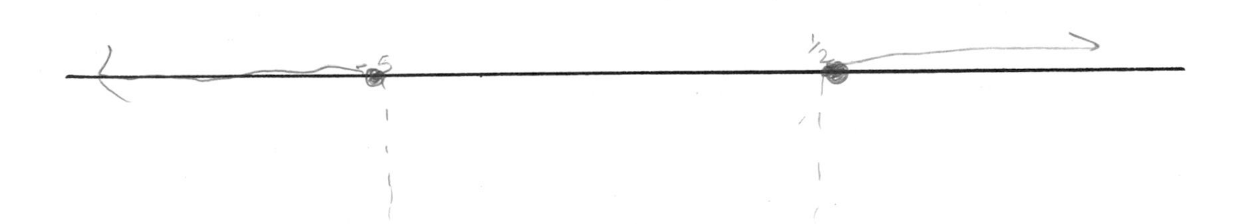

Test $x = -8$	Test $x = 0$	Test $x - 2$
$2(-8)^2+9(-8)-5>0$	$2(0)^2+9(0)-5>0$	$2(2)^2+9(2)-5>0$
51 ≥ 0 TRUE	−5 ≥ 0 FALSE	21 ≥ 0 TRUE
Region I does contain solutions.	Region II does not contain solutions.	Region III does contain solutions.

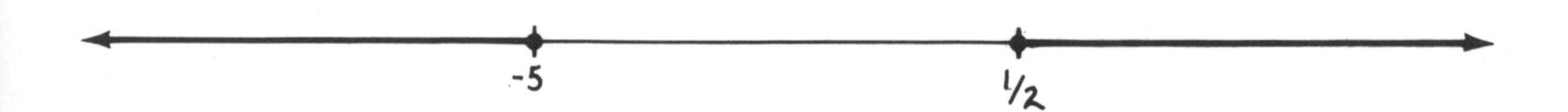

Note: -5 and 1/2 are solutions; therefore, they are denoted with a closed circle rather than an open circle.

Find the solutions of the following quadratic inequalities and graph the solutions on the number line. To save work, calculate the value of the discriminant first to determine if the solutions are imaginary.

2. $x^2 - 4x < 12$

$x^2 - 4x - 12 < 0$

$(x-6)(x+2) < 0$

$x = 6$ OR $x = -2$

I	-2	II	6	III
Test -4		Test 0		Test 7
$(-4)^2 - 4(-4) < 12$		$(0)^2 - 4(0) - 12 < 0$		$7^2 - 4(7) - 12 < 0$
$32 < 12$		$-12 < 0$		$9 < 0$
False		True		False

Solution:

3. $x^2 - 6x + 9 > 0$

$(x-3)^2 > 0$

$x = 3$

I	3	II
Test $x = 0$		Test $x = 4$
$0^2 - 6(0) + 9 > 0$		$4^2 - 6(4) + 9 > 0$
$9 > 0$		$1 > 0$
True		True

Solution:

3

Note: -5 and 1/2 are solutions; therefore, they are denoted with a closed circle rather than an open circle.

Find the solutions of the following quadratic inequalities and graph the solutions on the number line. To save work, calculate the value of the discriminant first to determine if the solutions are imaginary.

2. $x^2 - 4x < 12$

Solution: ______________________________

3. $x^2 - 6x + 9 > 0$

Solution: ______________________________

4. $3x^2 - 2x - 4 \leq 0$

$a = 3$

$b = -2$

$c = -4$

$x = \dfrac{2 \pm \sqrt{4+48}}{6}$

$= \dfrac{2 \pm 2\sqrt{13}}{6} = \dfrac{1 \pm \sqrt{13}}{3}$

$x \doteq 1.535$

$x \doteq -.869$

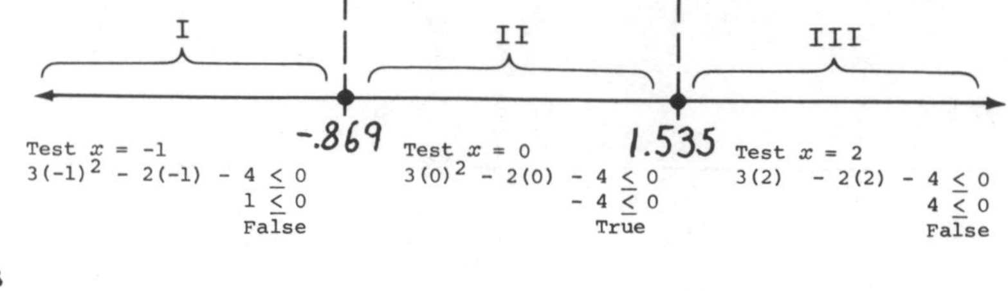

Solution:

5. $9x^2 - 16 \geq 0$

$(3x+4)(3x-4)=0$

$x = -\frac{4}{3}$ or $\frac{4}{3}$

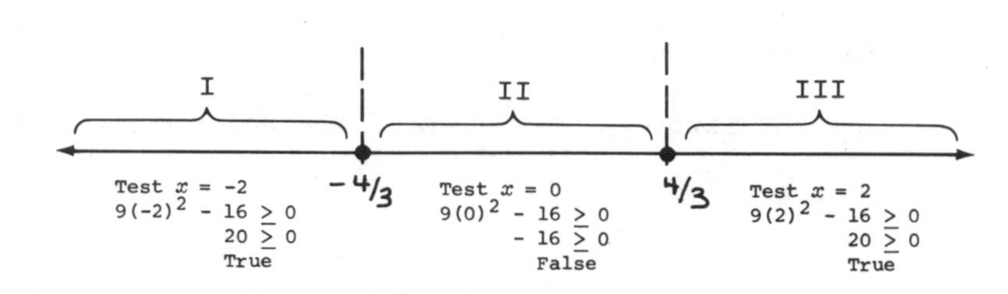

Solution:

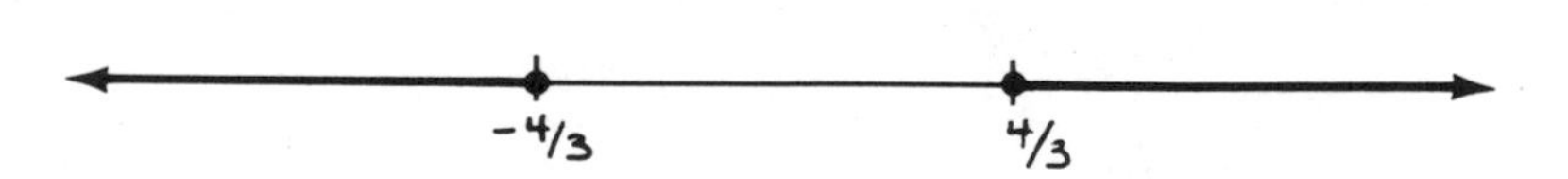

6. 67

$2x(x + 4) \leq x - 5$

$2x^2+8x \leq x-5$

$2x^2+7x+5 \leq 0$

$(2x+5)(x+1)=0$

$2x+5=0 \qquad x+1=0$

$x = -\frac{5}{2} \qquad x = -1$

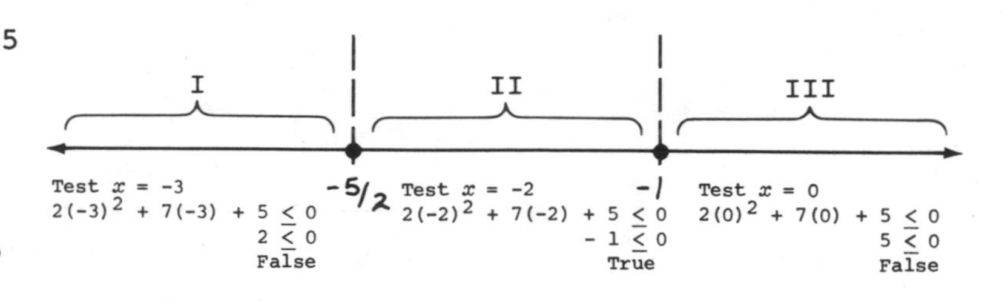

Solution:

4. $3x^2 - 2x - 4 \leq 0$

Solution:

5. $9x^2 - 16 \geq 0$

Solution:

6. 67 $2x(x + 4) \leq x - 5$

Solution:

7. $2x^2 + 4x \geq -3$

$2x^2 + 4x + 3 \geq 0$

$a = 2,\ b = 4,\ c = 3$

$x = \dfrac{-4 \pm \sqrt{16 - 24}}{4}$

Since the discriminant < 0, the roots are imaginary and cannot be graphed on the real number line.

Solution: ______________________

8. Given: $\dfrac{x + 1}{x - 3} \leq 0$

$x + 1 = 0$ $\quad$ $x - 3 = 0$ $\quad$ (1) Determine boundary points.

$x = \underline{-1}$ $\quad$ $x = \underline{3}$

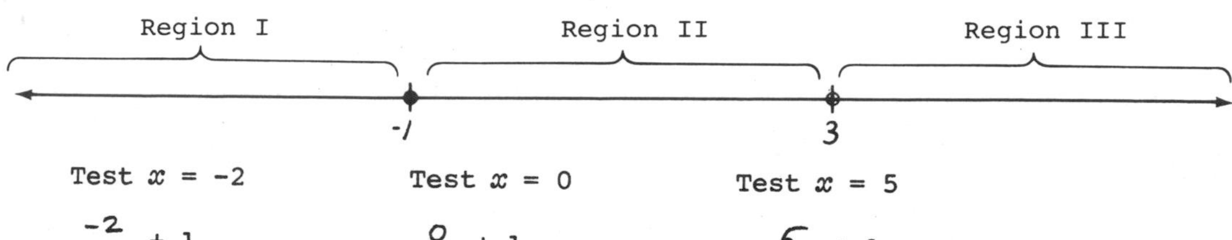

Test $x = -2$ $\quad$ Test $x = 0$ $\quad$ Test $x = 5$

$\dfrac{\underline{-2} + 1}{\underline{-2} - 3} \leq 0$ $\quad$ $\dfrac{\underline{0} + 1}{\underline{0} - 3} \leq 0$ $\quad$ $\dfrac{\underline{5} + 1}{\underline{5} - 3} \leq 0$

$\underline{\tfrac{1}{5}} \leq 0$ FALSE $\quad$ $\underline{-\tfrac{1}{3}} \leq 0$ TRUE $\quad$ $\underline{\tfrac{6}{2}} \leq 0$ FALSE

Region I does not contain solutions. $\quad$ Region II does contain solutions. $\quad$ Region III does not contain solutions.

Note: 3 is graphed with an open circle because if $x = 3$, the original inequality is undefined as there would be a zero in the denominator of a fraction.

7. $2x^2 + 4x \geq -3$

____

Solution: _______________________________

8. Given: $\frac{x + 1}{x - 3} \leq 0$

$x + 1 = 0$ $x - 3 = 0$ (1) Determine boundary points.

$x =$ ___ $x =$ ___

Test $x = -2$	Test $x = 0$	Test $x = 5$
$\frac{\_\_ + 1}{\_\_ - 3} \leq 0$	$\frac{\_\_ + 1}{\_\_ - 3} \leq 0$	$\frac{\_\_ + 1}{\_\_ - 3} \leq 0$
_____ ≤ 0 FALSE	_____ ≤ 0 TRUE	_____ ≤ 0 FALSE
Region I ____ ____ contain solutions.	Region II _____ contain solutions.	Region III ____ ____ contain solutions.

Note: 3 is graphed with an open circle because if $x = 3$, the original inequality is undefined as there would be a zero in the denominator of a fraction.

Find the boundary points of each inequality, test the regions formed by the boundary points, and graph the solutions on the number line.

9. $\dfrac{x}{x-5} > 2$

$\dfrac{x}{x-5} - 2 > 0$

$\dfrac{x}{x-5} - \dfrac{2(x-5)}{x-5} > 0$

$\dfrac{-x+10}{x-5} > 0$

$-x+10=0$ $\quad x=10$

$x-5=0$ $\quad x=5$

I	II	III
Test $x = 0$	Test $x = 7$	Test $x = 11$
$\dfrac{0}{0-5} > 2$	$\dfrac{7}{7-5} > 2$	$\dfrac{11}{11-5} > 2$
$0 > 2$	$\dfrac{7}{2} > 2$	$\dfrac{11}{6} > 2$
False	True	False

Boundary points on number line: 5, 10

Solution:

5 ——— 10

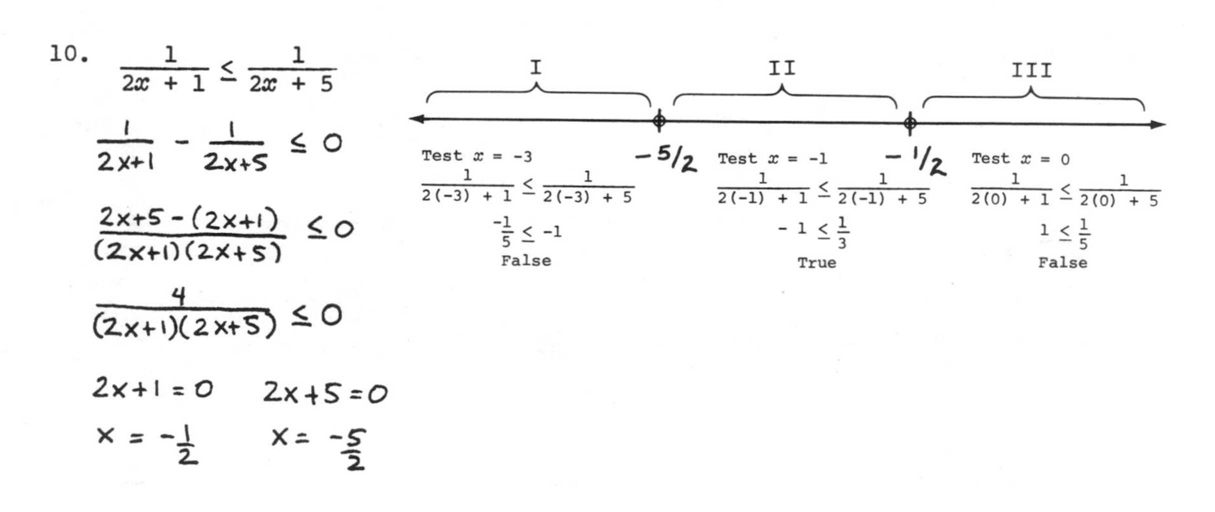

Solution:

−5/2 ——— −1/2

Find the boundary points of each inequality, test the regions formed by the boundary points, and graph the solutions on the number line.

9. $\dfrac{x}{x - 5} > 2$

Solution:

10. $\dfrac{1}{2x + 1} \le \dfrac{1}{2x + 5}$

Solution:

11. $(2x + 3)(x - 3)(x + 2) \geq 0$

$2x+3=0 \quad x-3=0 \quad x+2=0$

$x=-\frac{3}{2} \quad x=3 \quad x=-2$

Note: you can shorten your work time by determining the sign of each set of parentheses; you are simply checking to see if the product is ≥ 0.
Actual computation is not necessary.

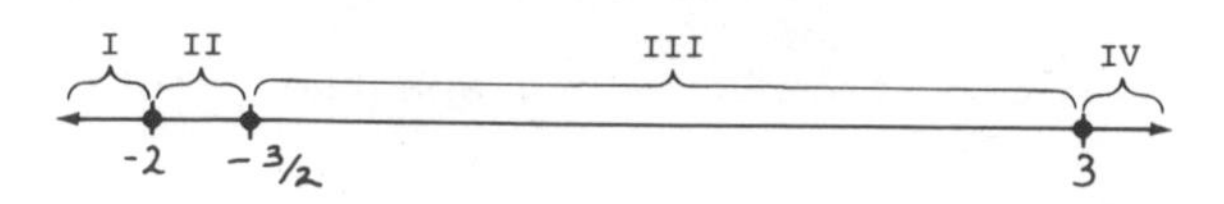

I: Test $x = -3$: $[2(-3) + 3][(-3) - 3][(-3) + 2] \geq 0$
$(-)(-)(-) \geq 0$ False

II: Test $x = \frac{-7}{4}$: $[2(\frac{-7}{4}) + 3][(\frac{-7}{4}) - 3][(\frac{-7}{4}) + 2] \geq 0$
$(-)(-)(-) \geq 0$ True

III: Test $x = 0$: $[2(0) + 3][(0) - 3][(0) + 2] \geq 0$
$(+)(-)(+) \geq 0$ False

IV: Test $x = 4$: $[2(4) + 3][(4) - 3][(4) + 2] \geq 0$
$(+)(+)(+) \geq 0$ True

Solution:

12. $x^3 + 3x^2 - 4x - 12 < 0$

$x^2(x+3)-4(x+3)<0$

$(x+3)(x^2-4)<0$

★ $(x+3)(x+2)(x-2)<0$

$x+3=0 \quad x+2=0$

$x-2=0$

$x=-3 \quad x=-2 \quad x=2$

I: Test $x = -4$: $(-4 + 3)(-4 + 2)(-4 - 2) < 0$
$(-)(-)(-) < 0$ True

II: Test $x = -2.5$: $(-2.5 + 3)(-2.5 + 2)(-2.5 - 2) < 0$
$(+)(-)(-) < 0$ False

III: Test $x = 0$: $(0 + 3)(0 + 2)(0 - 2) < 0$
$(+)(+)(-) < 0$ True

IV: Test $x = 4$: $(4 + 3)(4 + 2)(4 - 2) < 0$
$(+)(+)(+) < 0$ False

Note: Be sure your factored form is correct if you use it to test.
If you have doubt, use the original form of the equation.

Solution:

-3 -2 2

11. $(2x + 3)(x - 3)(x + 2) \geq 0$

____

Solution: ________________________________

12. $x^3 + 3x^2 - 4x - 12 < 0$

____

Solution: ________________________________

Chapter 7 Self-Test

Determine the vertex and sketch the graph of each function.

1. $y = 3x^2 + 4$

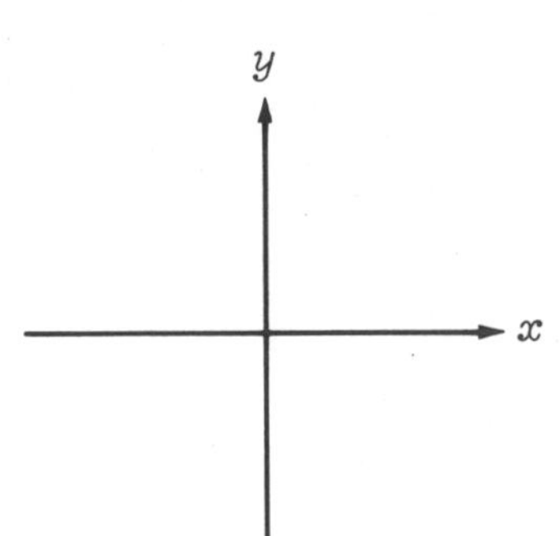

2. $y + 14 = 2x^2 + 12x$

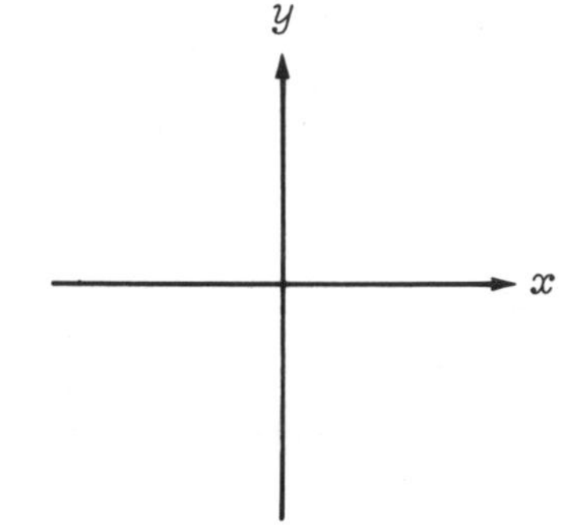

Solve the following quadratic equations by extracting the roots.

3. $2x^2 = 15$

4. $(x - 1)^2 + 72 = 0$

Solve the following quadratic equations by completing the square.

5. $x^2 - 14x - 51 = 0$

6. $3x^2 + 17x - 5 = 0$

Solve the following quadratic equations by use of the quadratic formula.

7. $3x = -4x^2 + 11$

8. $1/3 - 2/x - 1/x^2 = 0$

Solve each of the following:

9. The sum of a number and its reciprocal is 6. Find the number.

10. $2x^4 + x^2 - 3 = 0$

11. Find the solutions of $2x^2 - 15 < -7x$ and graph the solutions on a number line.

CHAPTER

8 Conic Sections

8·1 Horizontal Parabolas

1. A conic section is a curve which is formed by the intersection of a plane with a cone.

Match each item to its corresponding figure.

c 2. parabola

b 3. hyperbola

d 4. ellipse

a 5. circle

(a)

(b)

(c)

(d)

6. A vertical parabola is the graph of a function; a horizontal parabola is not a graph of a function because it does not satisfy the vertical line test.

C H A P T E R

8 Conic Sections

8 · 1 Horizontal Parabolas

1. A __________ ____________ is a curve which is formed by the intersection of a plane with a cone.

Match each item to its corresponding figure.

_____ 2. parabola

_____ 3. hyperbola

_____ 4. ellipse

_____ 5. circle

(a)

(b)

(c)

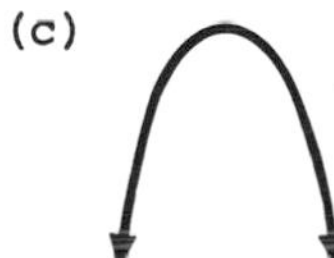

(d)

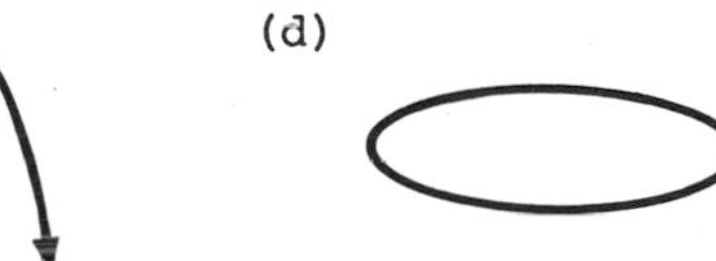

6. A _____________ parabola is the graph of a function; a ____________ parabola is not a graph of a function because it does not satisfy the __________________ test.

7. Given a quadratic equation in the form

$$x = a(y - k)^2 + h,\ a \neq 0$$

(a) The graph will be a horizontal parabola.

(b) If $a > 0$, the graph will open to the right; if $a < 0$, the graph will open to the left.

(c) The coordinates of the vertex will be (h, k).

(d) The horizontal line passing through the vertex is called the axis; its equation is $y = k$.

8. Given: $x = y^2 + 10y + 8$

$x = (y^2 + 10y \qquad) + 8$

$x = (y^2 + 10y + \underline{25}) + 8 - \underline{25}$ (1) $(1/2 \cdot b)^2 = (1/2 \cdot 10)^2 = 25$

$x = (\underline{y+5})^2 - 17$ (2) Factor.

The vertex is at $(-17, -5)$ and the parabola opens to the right.

Write the following quadratic equations in standard form. Determine the vertex and sketch the graph.

9. $x = -y^2 + 3y + 2$

$x = -(y^2 - 3y \quad) + 2$

$= -1(y^2 - 3y + \frac{9}{4}) + 2 + \frac{9}{4}$

$x = -1(y - \frac{3}{2})^2 + 4\frac{1}{4}$

vertex: $(h, k) = (4\frac{1}{4}, \frac{3}{2})$

axis: $y = \frac{3}{2}$

$a < 0$; opens to the left

x	y
$4\frac{1}{4}$	$\frac{3}{2}$
2	0
$3\frac{1}{2}$	$\frac{1}{4}$

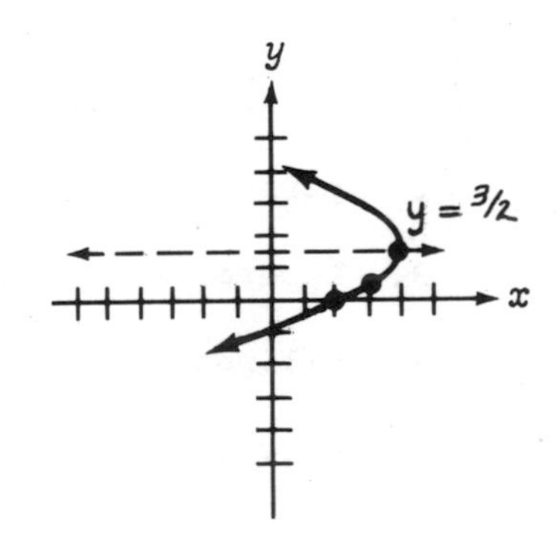

7. Given a quadratic equation in the form

$$x = a(y - k)^2 + h, \; a \neq 0$$

(a) The graph will be a ______________ ______________.

(b) If $a > 0$, the graph will open ________________ ; if $a < 0$, the graph will open ________________.

(c) The coordinates of the vertex will be __________.

(d) The horizontal line passing through the vertex is called the _________; its equation is __________.

8. Given: $x = y^2 + 10y + 8$

$x = (y^2 + 10y \quad\quad) + 8$

$x = (y^2 + 10y + \_\_\_) + 8 - \_\_\_$ (1) $(1/2 \cdot b)^2 = (1/2 \cdot 10)^2 = 25$

$x = (\_\_\_\_\_\_)^2 - 17$ (2) Factor.

The vertex is at __________ and the parabola opens to the _________.

Write the following quadratic equations in standard form. Determine the vertex and sketch the graph.

9. $x = -y^2 + 3y + 2$

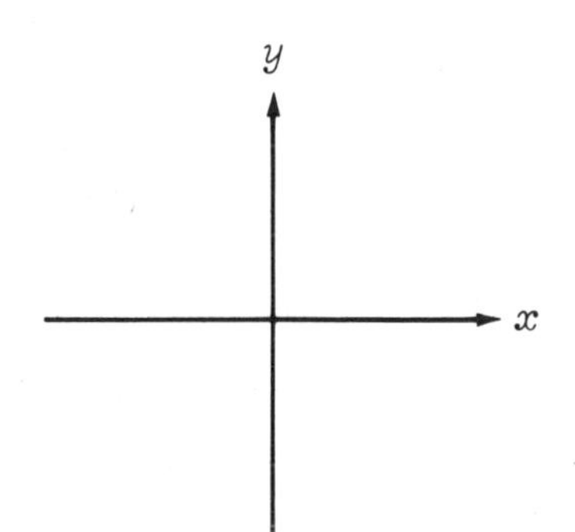

10. | 17

$x = y^2 - 3y + 1$

$x = (y^2 - 3y + \frac{9}{4}) + 1 - \frac{9}{4}$

$x = (y - \frac{3}{2})^2 - 1\frac{1}{4}$

vertex: $(-1\frac{1}{4}, \frac{3}{2})$

axis: $y = \frac{3}{2}$

$a > 0$; opens to the right

x	y
$-1\frac{1}{4}$	$\frac{3}{2}$
1	0
-1	2
$-\frac{1}{4}$	$\frac{1}{2}$

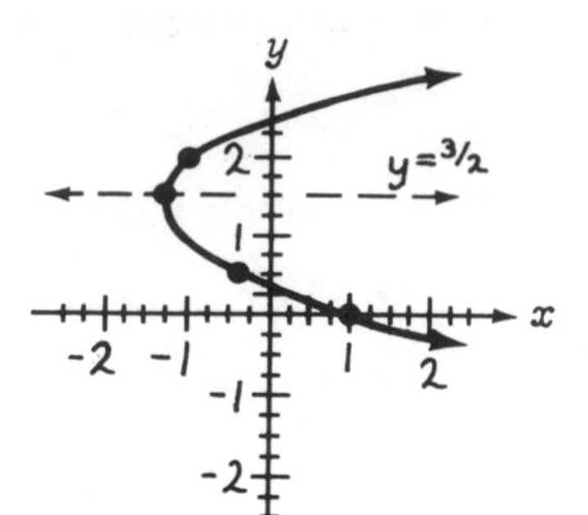

11. $x = 1/4\, y^2 + y + 5$

$x = \frac{1}{4}(y^2 + 4y \quad) + 5$

$= \frac{1}{4}(y^2 + 4y + 4) + 5 - 1$

$x = \frac{1}{4}(y + 2)^2 + 4$

vertex: $(4, -2)$

axis: $y = -2$

x	y
4	-2
5	0
5	-4

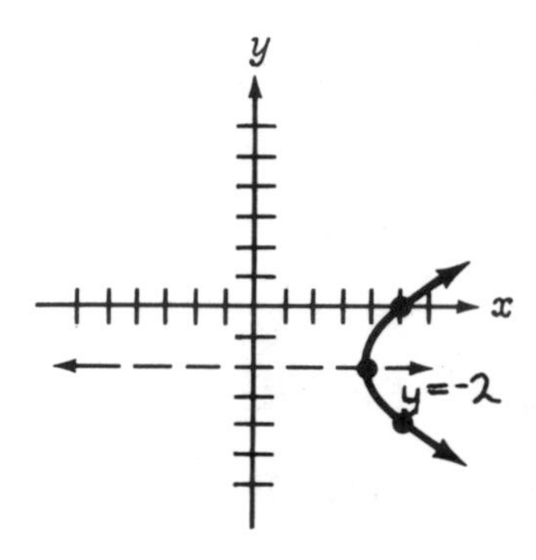

12. | 11

$x = -(y + 4)^2 - 2$

vertex: $(-2, -4)$

axis: $y = -4$

x	y
-2	-4
-3	-3
-3	-5

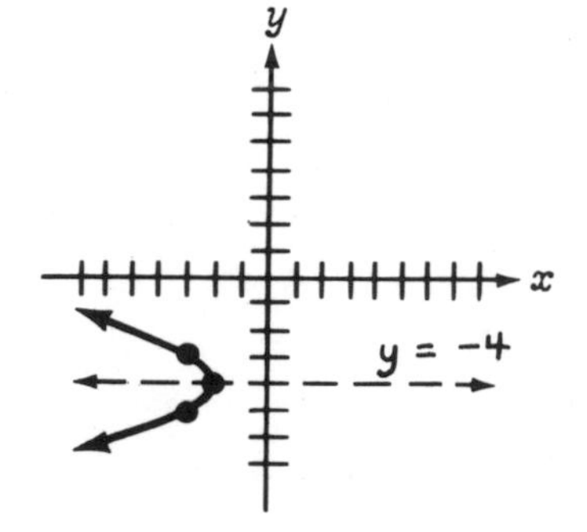

10.
17

$x = y^2 - 3y + 1$

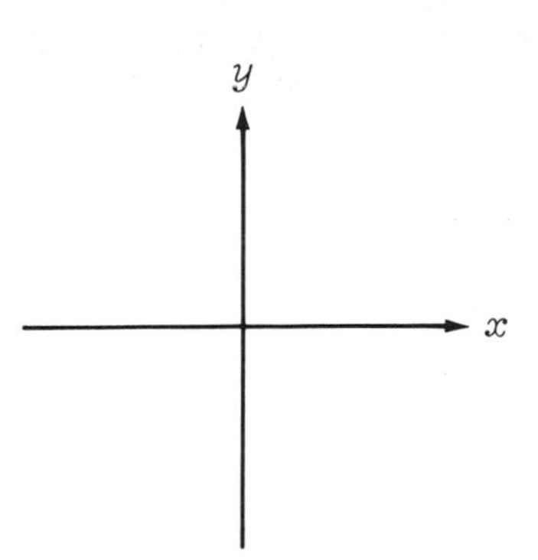

11. $x = 1/4\ y^2 + y + 5$

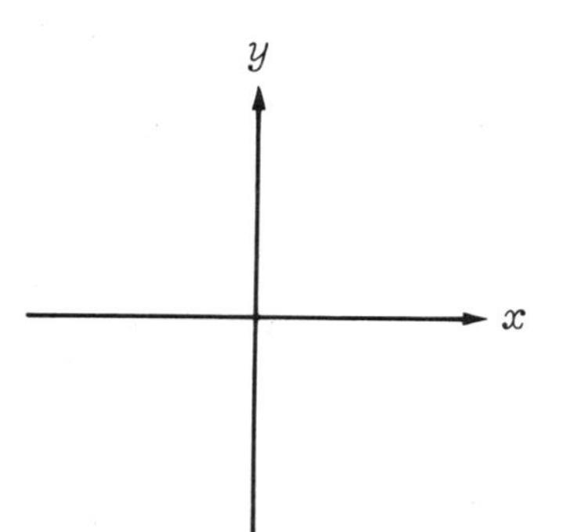

12.
11

$x = -(y + 4)^2 - 2$

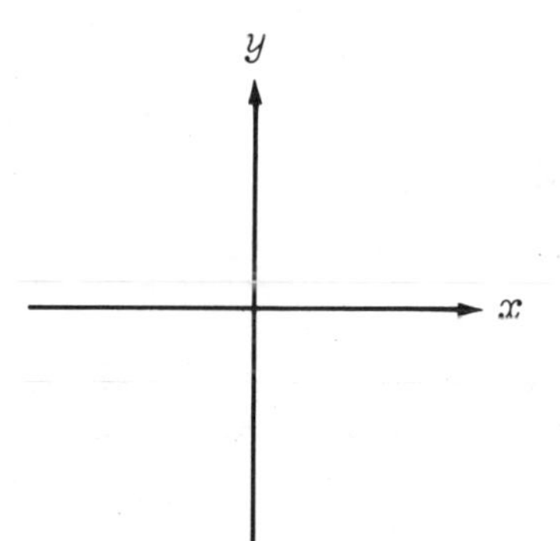

8 · 2 Circles

1. A circle is a set of points in a plane which are located a fixed distance, called the radius, from a given point in the plane, called the center. Note that the equation of a circle is not a function because its graph does not pass the vertical line test.

2. The equation of a circle is $(x-h)^2+(y-k)^2=r^2$. The coordinates of the center are (h,k) and the length of the radius is r.

3. Given: $x^2 + y^2 = 4.$

   $(x - 0)^2 + (y-0)^2 = 4$ (1) Put in standard form.

   $h = 0$, $k = 0$ $r = 2$

   The center of the circle is at $(0,0)$. The radius is 2.

4. Given: $x^2 - 2x + y^2 + 6y = 26$

   $(x^2 - 2x + 1) + (y^2 + 6y + 9) = 26 + 1 + 9$

   $(x-1)^2 + (y+3)^2 = 36$

   $h = 1$ $k = -3$ $r = 6$

   The center of the circle is at $(1,-3)$. The radius is 6.

Determine the equation of each of the following circles and graph.

5. Center at $(1/2, -2)$; radius = 3

   $h = \frac{1}{2}$, $k = -2$, $r = 3$

   $(x-\frac{1}{2})^2 + (y+2)^2 = 3^2$

   $x^2 - x + \frac{1}{4} + y^2 + 4y + 4 = 9$

   $x^2 + y^2 - x + 4y = \frac{19}{4}$

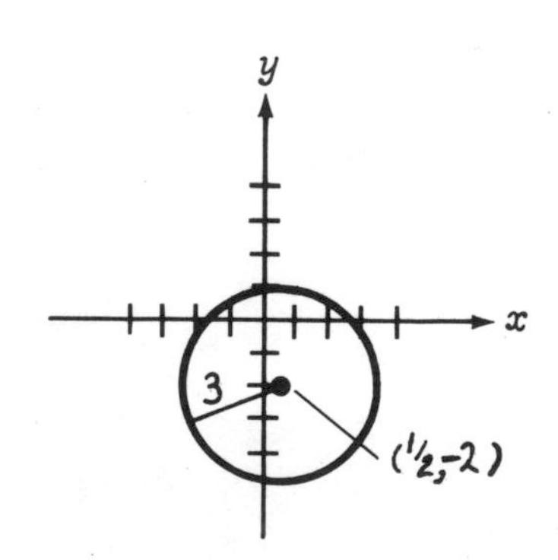

8·2 Circles

1. A ____________ is a set of points in a plane which are located a fixed distance, called the ____________, from a given point in the plane, called the ____________. Note that the equation of a circle is not a function because its graph does not pass the vertical line test.

2. The equation of a circle is ________________________________. The coordinates of the center are ___________ and the length of the radius is _____.

3. Given: $x^2 + y^2 = 4.$

   $(x - 0)^2 + (\_\_\_\_\_\_\_)^2 = 4$ (1) Put in standard form.

   $h =$ _____ $k =$ _____ $r =$ _____

   The center of the circle is at _________. The radius is ________.

4. Given: $x^2 - 2x + y^2 + 6y = 26$

   $(x^2 - 2x + \_\_\_) + (y^2 + 6y + \_\_\_) = 26 + \_\_\_ + \_\_\_$

   $(\_\_\_\_\_\_\_\_\_)^2 + (\_\_\_\_\_\_\_\_\_)^2 =$ ___________

   $h =$ _______ $k =$ _______ $r =$ _______

   The center of the circle is at _______. The radius is ______.

Determine the equation of each of the following circles and graph.

5. | 3 | Center at (1/2, -2); radius = 3

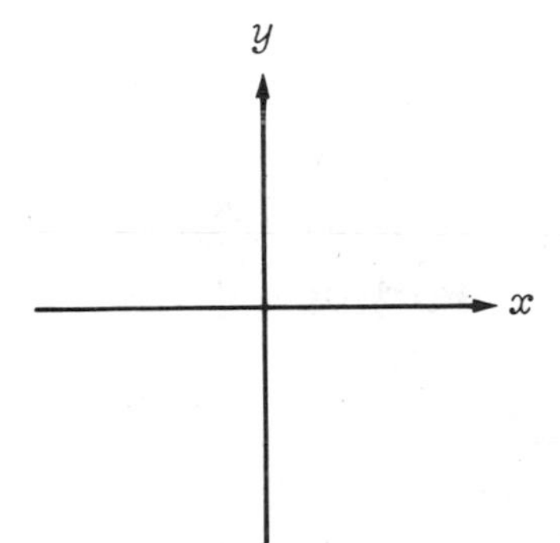

6. $(x - 4)^2 + (y + 2)^2 = 9$

$x^2 - 8x + 16 + y^2 + 4y + 4 = 9$

$x^2 + y^2 - 8x + 4y = -11$

center: $(h, k) : (4, -2)$

radius: 3

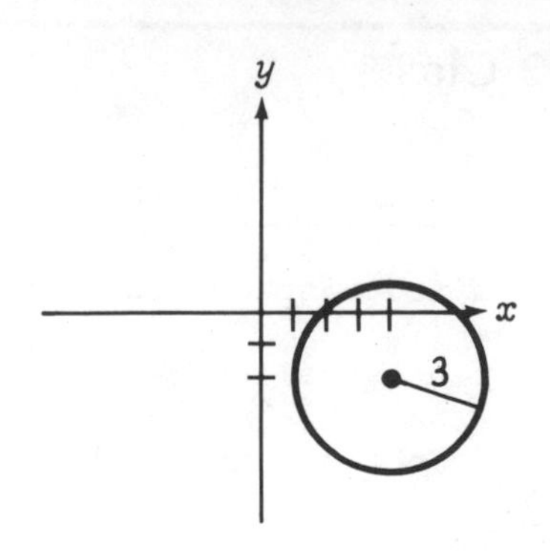

7. $x^2 + y^2 + 4x - 8y - 5 = 0$

$(x^2 + 4x + \quad) + (y^2 - 8y + \quad) = 5$

$(x^2 + 4x + 4) + (y^2 - 8y + 16) = 5 + 4 + 16$

$(x+2)^2 + (y-4)^2 = 25$

$h = -2$, $k = 4$, $r = 5$

center: $(-2, 4)$

radius: 5

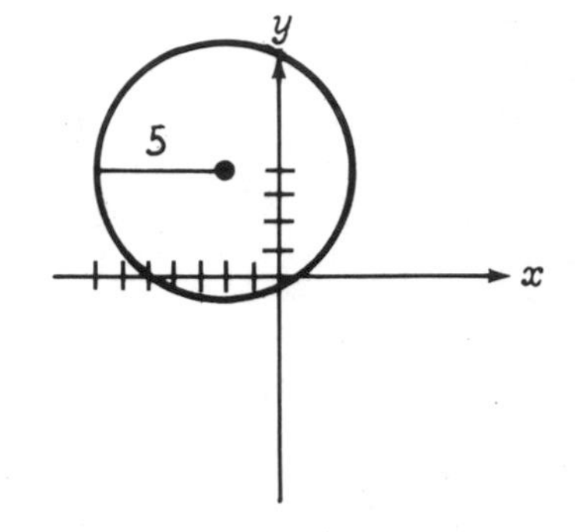

8. $x^2 + y^2 - 9x = -17/4$

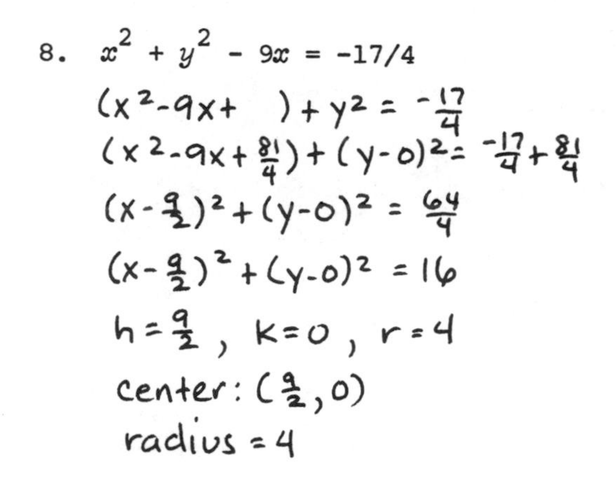

$(x^2 - 9x + \quad) + y^2 = -\frac{17}{4}$

$(x^2 - 9x + \frac{81}{4}) + (y-0)^2 = -\frac{17}{4} + \frac{81}{4}$

$(x - \frac{9}{2})^2 + (y-0)^2 = \frac{64}{4}$

$(x - \frac{9}{2})^2 + (y-0)^2 = 16$

$h = \frac{9}{2}$, $k = 0$, $r = 4$

center: $(\frac{9}{2}, 0)$

radius $= 4$

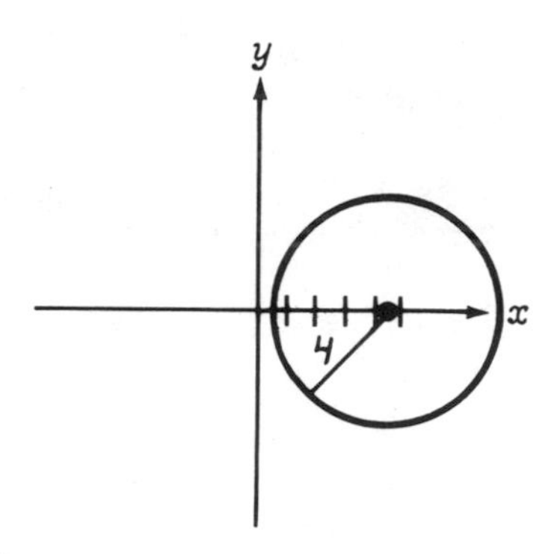

6. $(x - 4)^2 + (y + 2)^2 = 9$

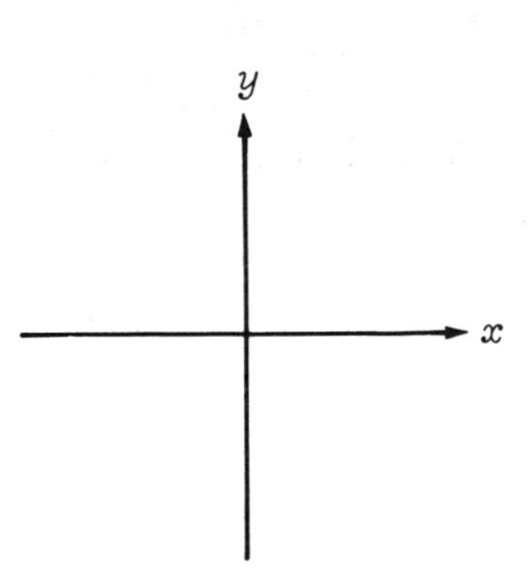

7. $x^2 + y^2 + 4x - 8y - 5 = 0$

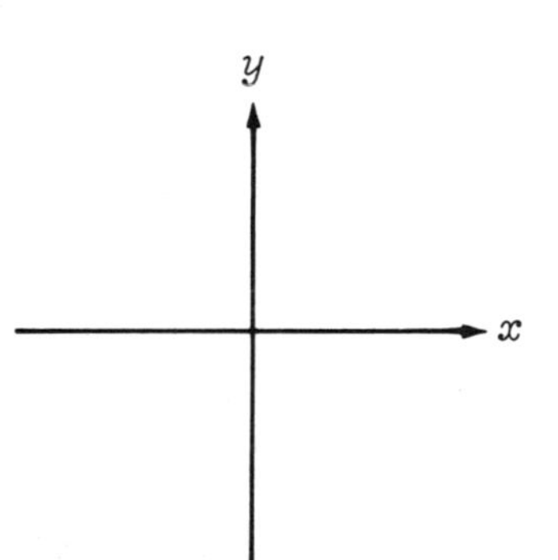

8. $x^2 + y^2 - 9x = -17/4$

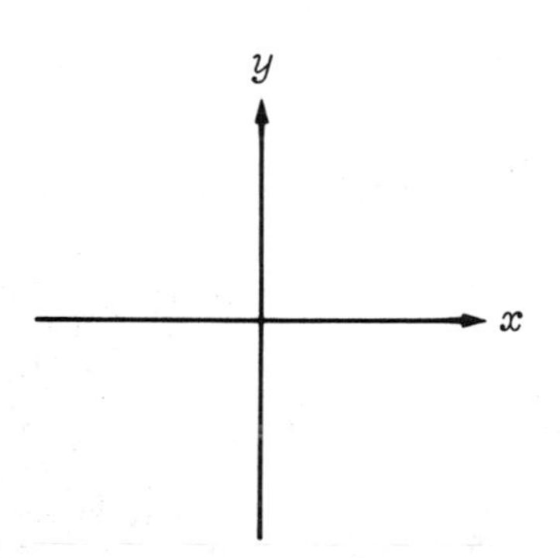

9. 25 $x^2 + y^2 + 2/3\ y = 0$

$x^2 + (y^2 + \frac{2}{3}y + \frac{1}{9}) = 0 + \frac{1}{9}$

$(x+0)^2 + (y+\frac{1}{3})^2 = \frac{1}{9}$

$h = 0,\ k = -\frac{1}{3},\ r = \frac{1}{3}$

center : $(0, -\frac{1}{3})$

radius : $\frac{1}{3}$

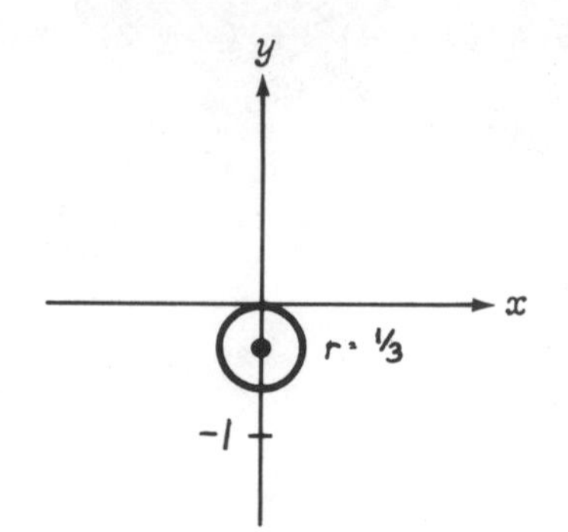

8 · 3 Ellipses

1. An ellipse is the set of all points in a plane the sum of whose distances from two fixed points in the plane, called foci, is constant. One "foci" is called a focus.

2. The general form of the equation of an ellipse with the center at (0, 0) is $\frac{x^2}{a^2} + \frac{y^2}{b^2} = 1$. The x-intercepts (when $y = 0$) are at $(a, 0)$ and $(-a, 0)$. The y-intercepts (when $x = 0$) are at $(0, b)$ and $(0, -b)$.

3. The line segments passing through the center and joining the intercepts are called the axes. If $a > b$, then the segment between $(a, 0)$ and $(-a, 0)$ is called the major axis. The segment between $(0, b)$ and $(0, -b)$ is called the minor axis.

4. The general form of the equation of an ellipse with the center at (h, k) is $\frac{(x-h)^2}{a^2} + \frac{(y-k)^2}{b^2} = 1$.

9. 25 $x^2 + y^2 + 2/3\ y = 0$

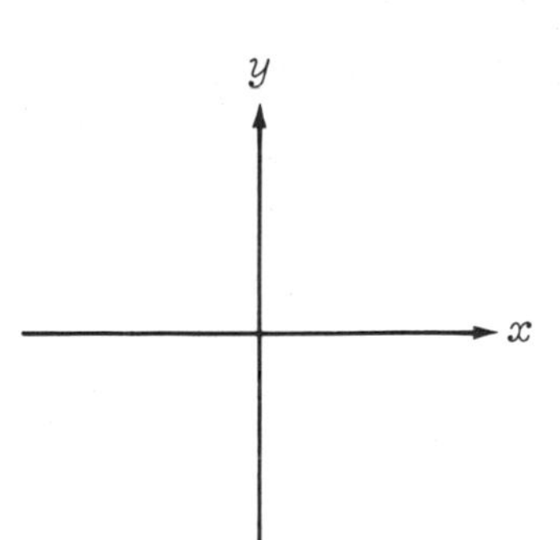

8·3 Ellipses

1. An ____________ is the set of all points in a plane the sum of whose ______________ from two fixed points in the plane, called __________, is constant. One "foci" is called a __________.

2. The general form of the equation of an ellipse with the center at (0, 0) is ________________. The x-intercepts (when $y = 0$) are at ____________ and ____________. The y-intercepts (when $x = 0$) are at ____________ and ____________.

3. The line segments passing through the center and joining the intercepts are called the _________. If $a > b$, then the segment between $(a, 0)$ and $(-a, 0)$ is called the __________ __________. The segment between $(0, b)$ and $(0, -b)$ is called the __________ __________.

4. The general form of the equation of an ellipse with the center at (h, k) is ____________________________.

Graph the following equations. Determine the center, the x-intercepts, and the y-intercepts.

5. $\dfrac{x^2}{36} + \dfrac{y^2}{4} = 1$

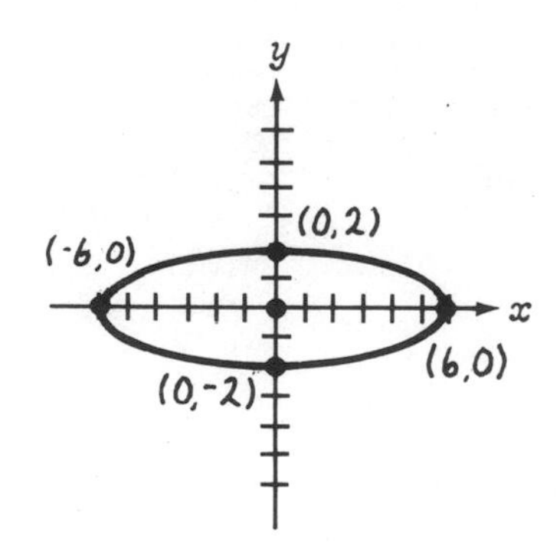

center: $(0,0)$

x-intercepts: $(6,0)$ and $(-6,0)$

y-intercepts: $(0,2)$ and $(0,-2)$

Note: If the center is not $(0,0)$, the points corresponding to the x and y intercepts will be called critical points in future problems.

6. / 13. $\dfrac{(x + 1/2)^2}{9} + \dfrac{(y + 4)^2}{5} = 1$

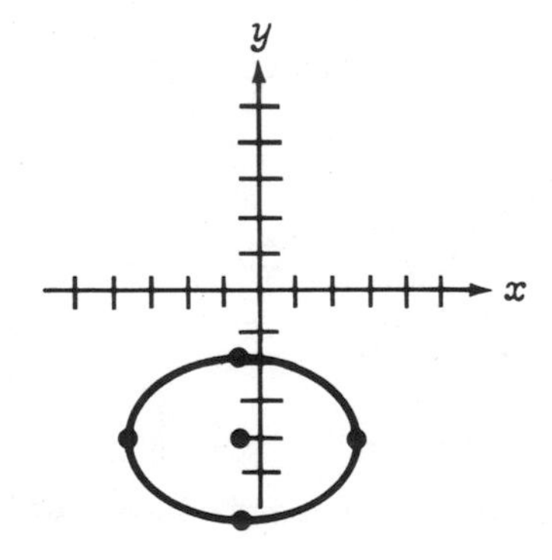

Center: $(-\frac{1}{2}, -4)$

Critical points: $(-\frac{1}{2}+3, -4)$, $(-\frac{1}{2}-3, -4)$

critical points: $(-\frac{1}{2}, -4+\sqrt{5})$, $(-\frac{1}{2}, -4-\sqrt{5})$

7. $\dfrac{12x^2}{48} + \dfrac{3y^2}{48} = \dfrac{48}{48}$

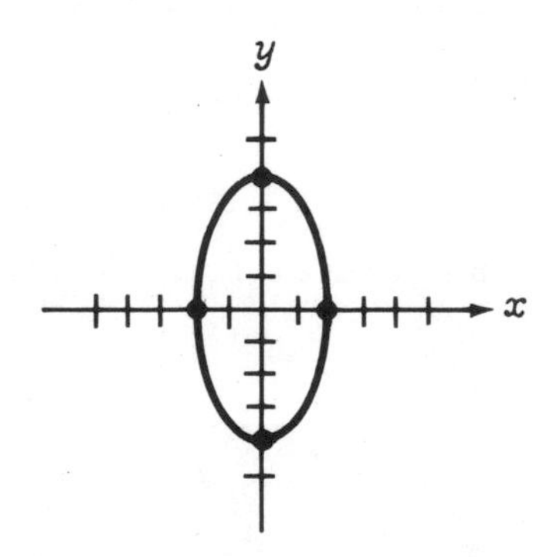

$\dfrac{x^2}{4} + \dfrac{y^2}{16} = 1$

Center: $(0,0)$

x-intercepts: $(2,0), (-2,0)$

y-intercepts: $(0,4), (0,-4)$

Graph the following equations. Determine the center, the x-intercepts, and the y-intercepts.

5. $\frac{x^2}{36} + \frac{y^2}{4} = 1$

center:

x-intercepts:

y-intercepts:

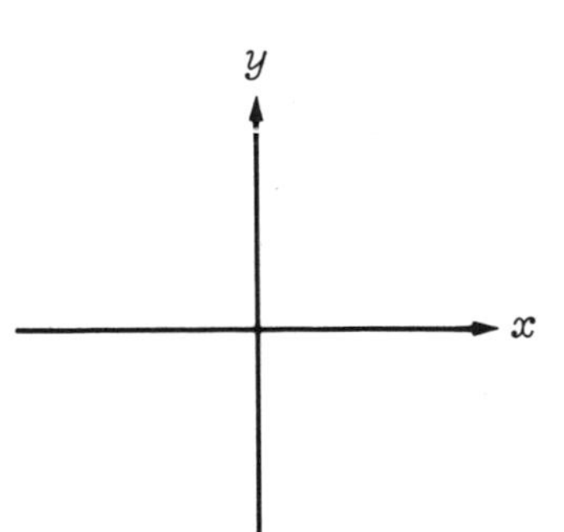

6. | 13 $\frac{(x + 1/2)^2}{9} + \frac{(y + 4)^2}{5} = 1$

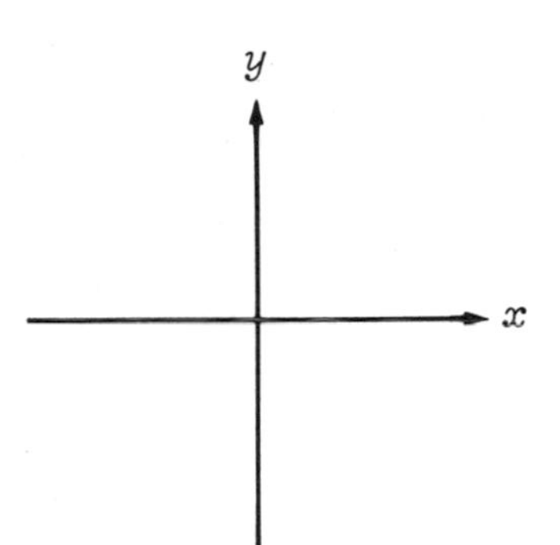

7. $12x^2 + 3y^2 = 48$

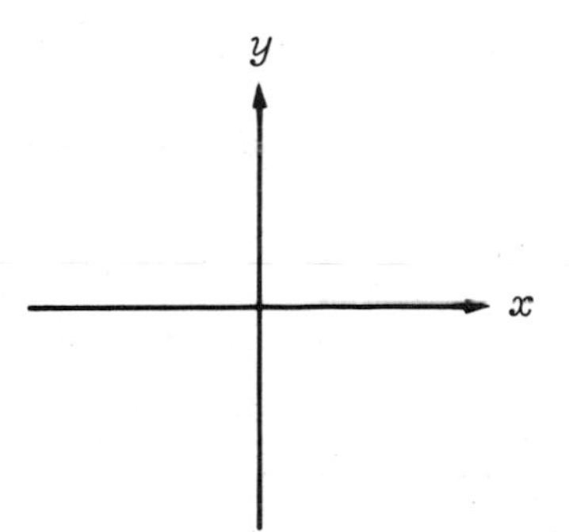

8. 17

$9x^2 + 25y^2 - 18x - 150y + 9 = 0$

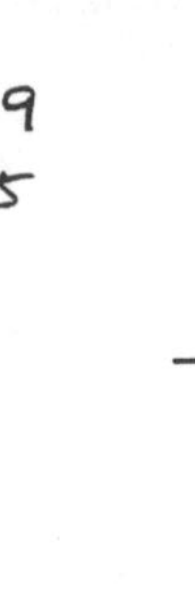

$(9x^2 - 18x + \quad) + (25y^2 - 150y \quad) = -9$

$9(x^2 - 2x + 1) + 25(y^2 - 6y + 9) = -9 + 9 + 225$

$\frac{9(x-1)^2}{225} + \frac{25(y-3)^2}{225} = \frac{225}{225}$

$\frac{(x-1)^2}{25} + \frac{(y-3)^2}{9} = 1$

Center: (1, 3)

Critical points: (-4,3), (6,3)

Critical points: (1,0), (1,6)

---

8 . 4 Hyperbolas

1. A hyperbola is the set of all points in a plane the difference of whose distances from two fixed points called the foci is constant. The midpoint of the segment joining the foci is called the center.

2. The general form of the equation of a hyperbola with center at (0, 0) is $\frac{x^2}{a^2} - \frac{y^2}{b^2} = 1$. This equation of a hyperbola has no y-intercepts. The x-intercepts (when $y = 0$) are at $(a, 0)$ and $(-a, 0)$. The x-intercepts are called the vertices of the hyperbola.

3. Given: $\frac{x^2}{9} - \frac{y^2}{4} = 1$

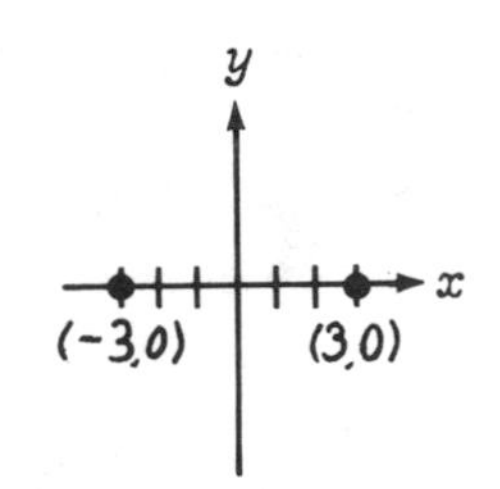

(a) Plot the vertices (x-intercepts).

(3,0) and (-3,0)

8. /7 $9x^2 + 25y^2 - 18x - 150y + 9 = 0$

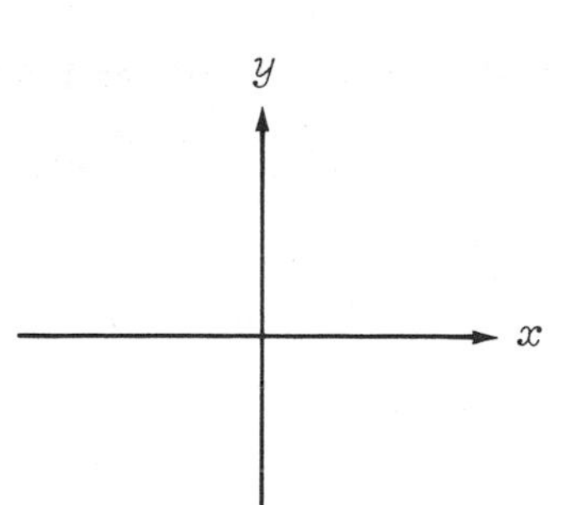

8·4 Hyperbolas

1. A ______________ is the set of all points in a plane the difference of whose distances from two fixed points called the __________ is constant. The midpoint of the segment joining the foci is called the ____________.

2. The general form of the equation of a hyperbola with center at (0, 0) is ________________. This equation of a hyperbola has no ___-intercepts. The x-intercepts (when $y = 0$) are at ________ and ________. The x-intercepts are called the ______________ of the hyperbola.

3. Given: $\frac{x^2}{9} - \frac{y^2}{4} = 1$

   (a) Plot the vertices (x-intercepts).

   _______ and _______

(b) Plot $(0, b)$ and $(0, -b)$.

(0,2) and (0,-2)

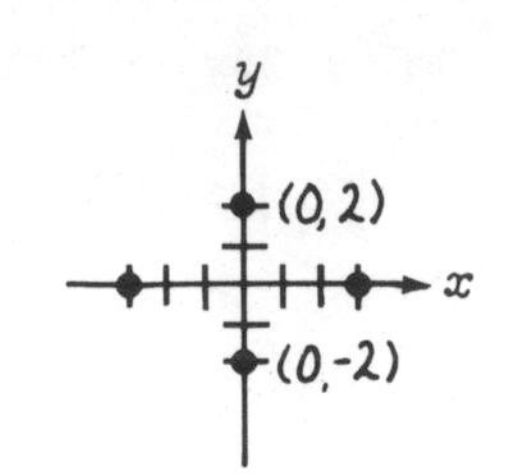

(c) Sketch a rectangle with sides parallel to the x- and y-axes passing through these points.

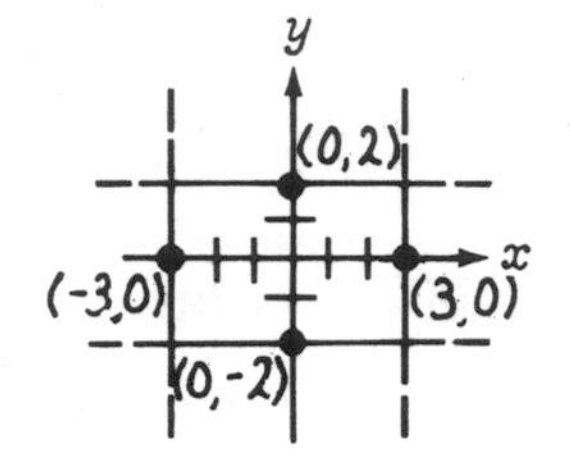

(d) Draw the extended diagonals of the rectangle. These diagonals are called the asymptotes of the hyperbola.

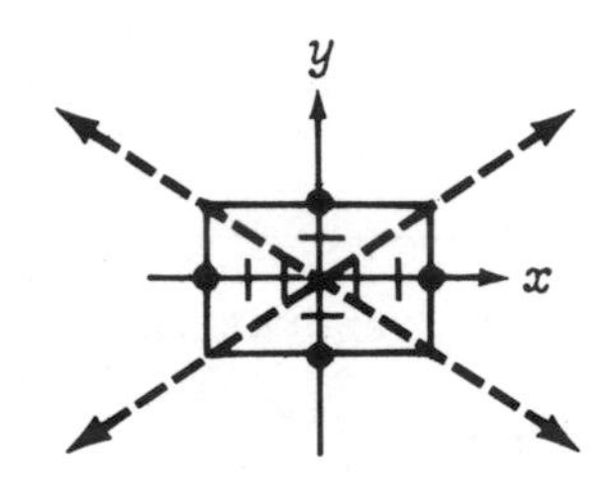

(e) Sketch the graph of the hyperbola. Note that the graph approaches the asymptotes without ever touching them and is symmetrical with respect to the x- and y-axes.

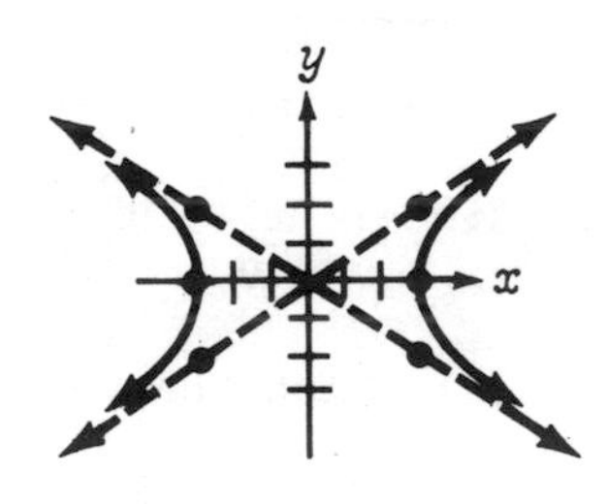

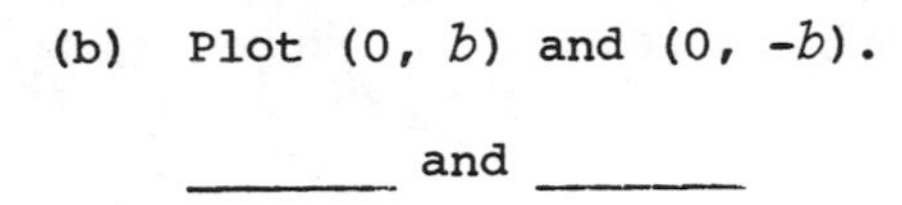

(b) Plot $(0, b)$ and $(0, -b)$.

_______ and _______

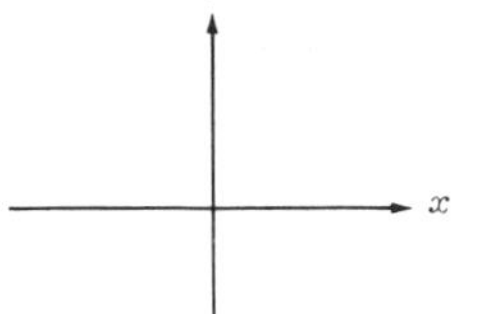

(c) Sketch a rectangle with sides parallel to the x- and y-axes passing through these points.

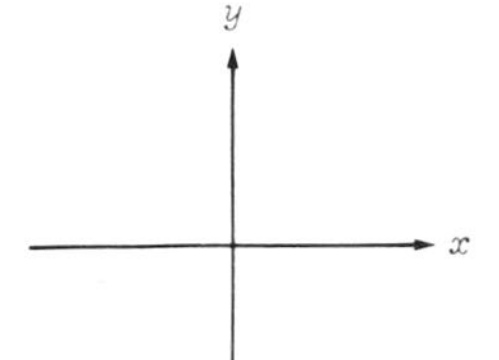

(d) Draw the extended diagonals of the rectangle. These diagonals are called the ________________ of the hyperbola.

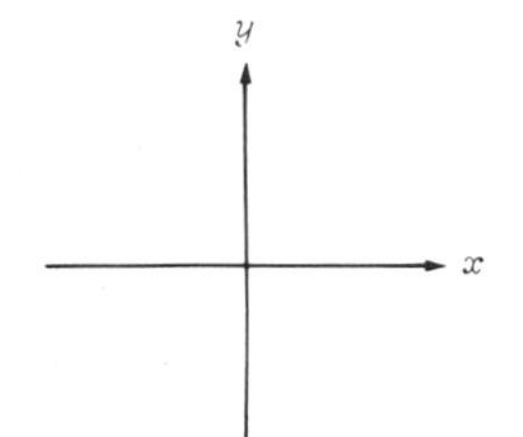

(e) Sketch the graph of the hyperbola. Note that the graph approaches the asymptotes without ever touching them and is symmetrical with respect to the x- and y-axes.

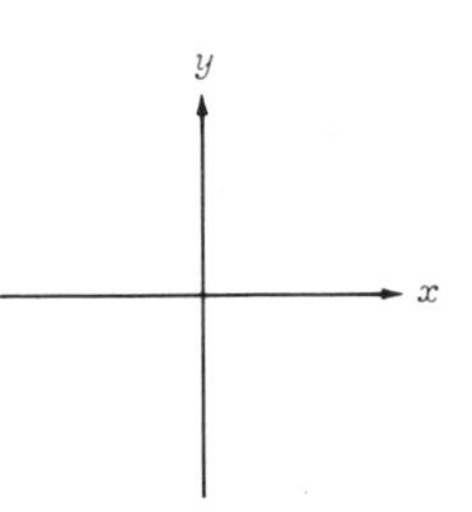

4. If the general form of the equation of a hyperbola is expressed in the form

$$\frac{y^2}{a^2} - \frac{x^2}{b^2} = 1$$

then this equation has no x-intercepts. The y-intercepts are at $(0, a)$ and $(0, -a)$.

5. The general form of the equation of a hyperbola with center at (h, k) is $\frac{(x-h)^2}{a^2} - \frac{(x-k)^2}{b^2} = 1$.

Graph the following equations. Determine the vertices and the asymptotes.

6. $\frac{y^2}{9} - \frac{x^2}{9} = \frac{9}{9}$ $\quad \frac{y^2}{9} - \frac{x^2}{9} = 1$

vertices: $(0, 3)$ and $(0, -3)$

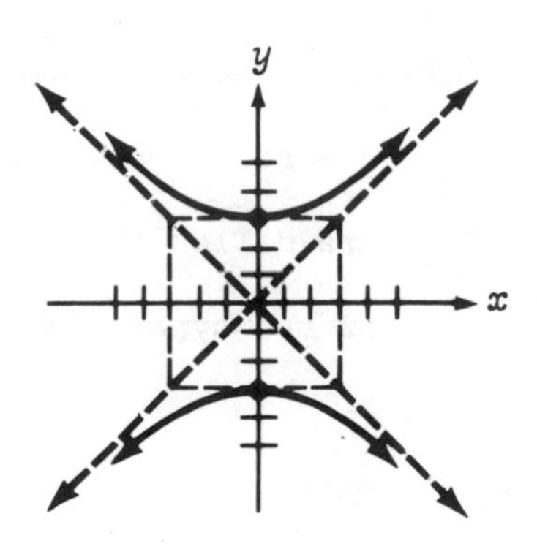

7. $\frac{9y^2}{-18} - \frac{2x^2}{-18} = \frac{-18}{-18}$

9

$$\frac{-y^2}{2} + \frac{x^2}{9} = 1$$

$$\frac{x^2}{9} - \frac{y^2}{2} = 1$$

vertices: $(3, 0)$ and $(-3, 0)$

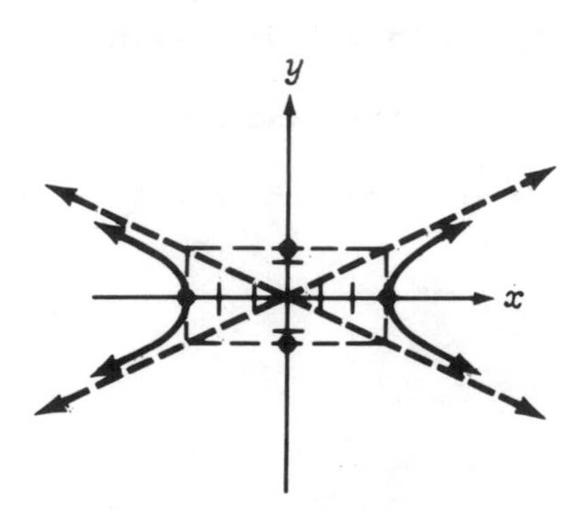

4. If the general form of the equation of a hyperbola is expressed in the form

$$\frac{y^2}{a^2} - \frac{x^2}{b^2} = 1$$

then this equation has no ___-intercepts. The y-intercepts are at _______ and _______.

5. The general form of the equation of a hyperbola with center at (h, k) is ___________________________.

Graph the following equations. Determine the vertices and the asymptotes.

6. $y^2 - x^2 = 9$

vertices:

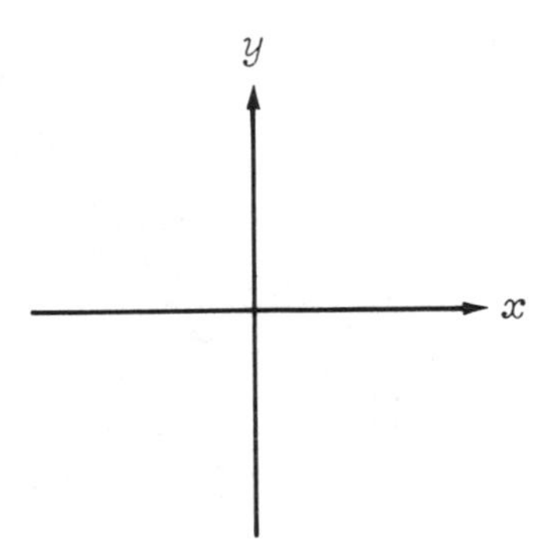

7. | 9 | $9y^2 - 2x^2 = -18$

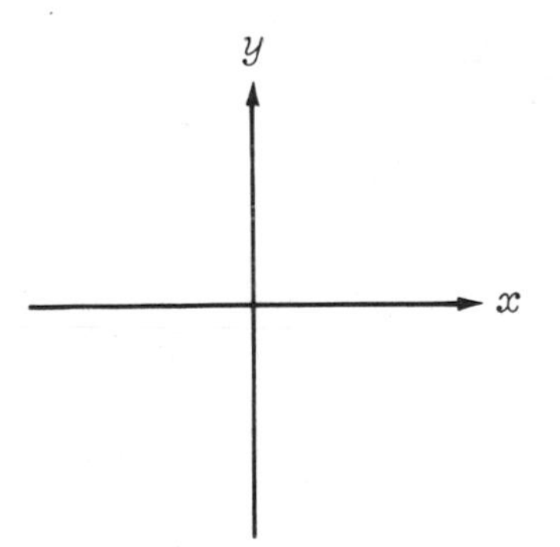

8. $\dfrac{(x+2)^2}{9} - \dfrac{(y-1)^2}{16} = 1$

center: $(-2,1)$

vertices: $(-2+3,1)$ and $(-2-3,1)$

$(1,1)$ and $(-5,1)$

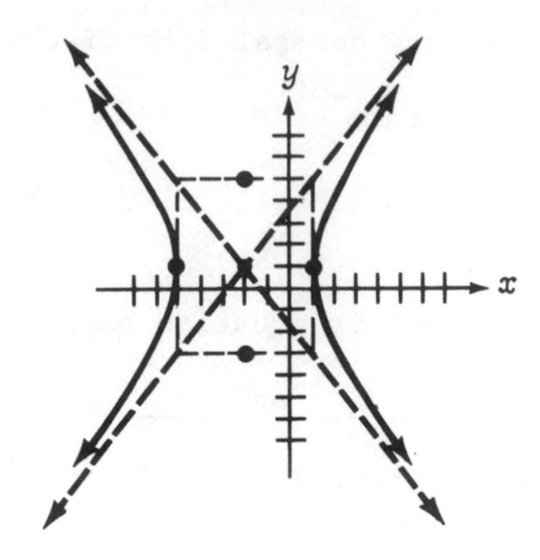

9. / 19 $x^2 - 36y^2 - 2x + 288y - 539 = 0$

$(x^2-2x\quad) - 36(y^2-8y\quad) = 539$

$(x^2-2x+1) - 36(y^2-8y+16) = 539 + 1 - 576$

$(x-1)^2 - 36(y-4)^2 = -36$

$\dfrac{(y-4)^2}{1} - \dfrac{(x-1)^2}{36} = 1$

Center: $(4,1)$

vertices: $(4,0)$ and $(4,2)$

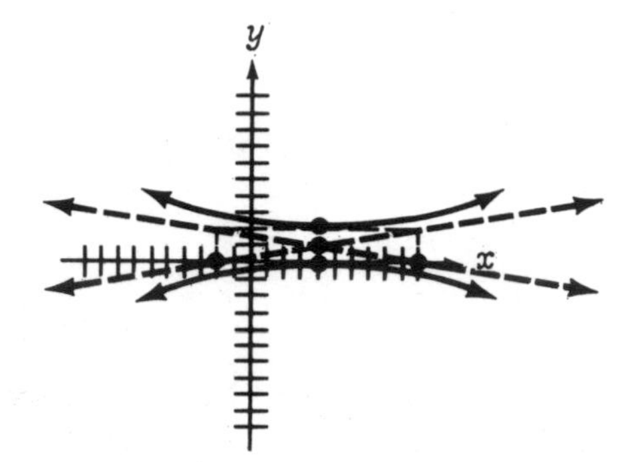

10. $\dfrac{y^2}{4/9} - \dfrac{x^2}{1/25} = 1$

center: $(0,0)$

vertices: $(0,\frac{2}{3}),(0,-\frac{2}{3})$

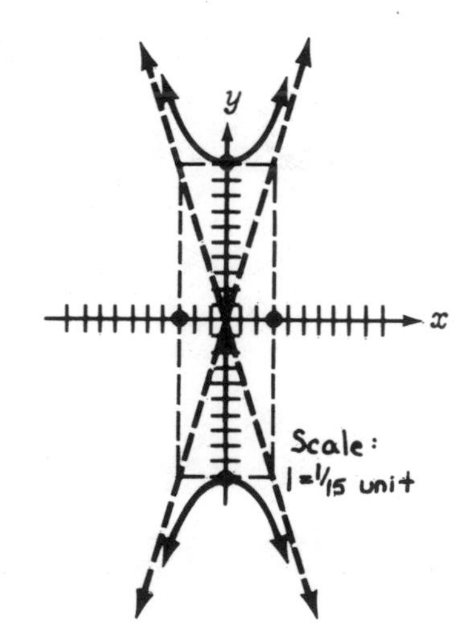

8. $\dfrac{(x + 2)^2}{9} - \dfrac{(y - 1)^2}{16} = 1$

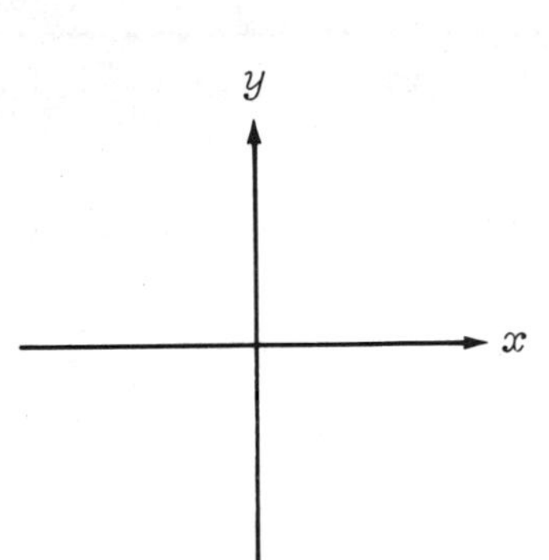

9.
19

$x^2 - 36y^2 - 2x + 288y - 539 = 0$

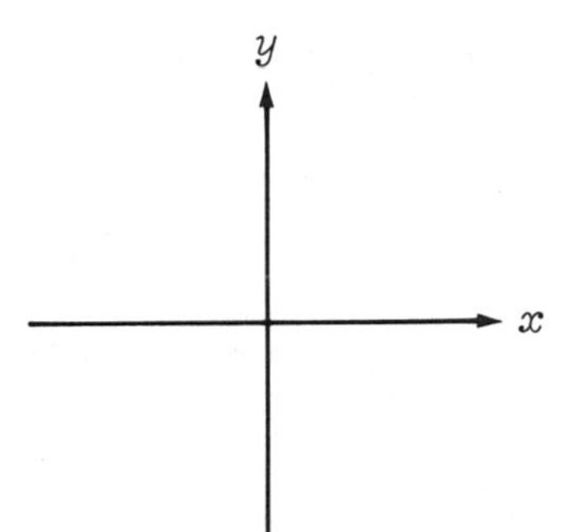

10. $\dfrac{y^2}{4/9} - \dfrac{x^2}{1/25} = 1$

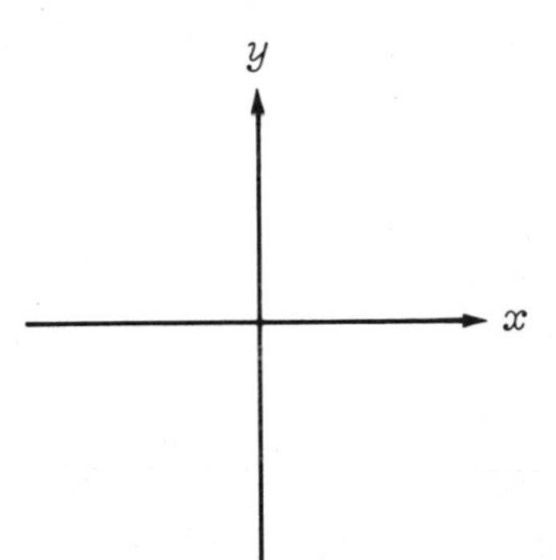

8 · 5 Graphing Second-Degree Inequalities

Each equation below describes a conic section. Write circle, ellipse, hyperbola, or parabola in the blank.

circle 1. $x^2 + y^2 = 9$

parabola 2. $y = x^2 + 4x + 4$

hyperbola 3. $2x^2 - 2y^2 = 16$

ellipse 4. $2x^2 + y^2 + 16x - 2y = -1$

hyperbola 5. $3x^2 - y^2 = -15$

ellipse 6. $(x + 3)^2 + (y - 2)^2 = 5$

parabola 7. $x = 2y^2 + 3y + 10$

ellipse 8. $3x^2 + y^2 = 9$

Graph the following inequalities. Sketch the conic section and shade the appropriate region.

9. $x^2 + y^2 < 16$

conic section: Circle

center: (0, 0)

radius: 4

Test (0,0): $0^2+0^2<16$ True

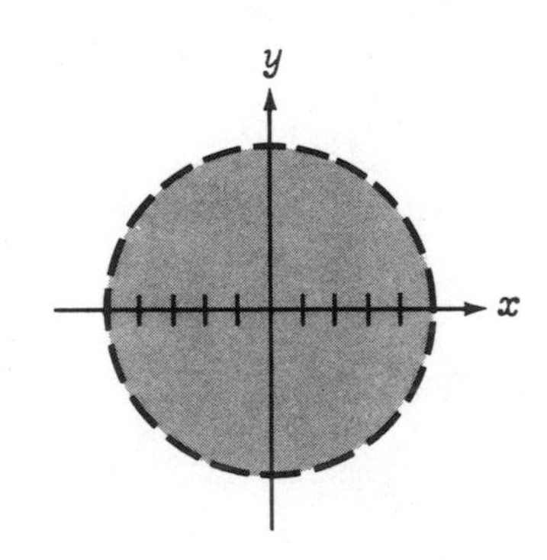

8 · 5 Graphing Second-Degree Inequalities

Each equation below describes a conic section. Write circle, ellipse, hyperbola, or parabola in the blank.

______________ 1. $x^2 + y^2 = 9$

______________ 2. $y = x^2 + 4x + 4$

______________ 3. $2x^2 - 2y^2 = 16$

______________ 4. $2x^2 + y^2 + 16x - 2y = -1$

______________ 5. $3x^2 - y^2 = -15$

______________ 6. $(x + 3)^2 + (y - 2)^2 = 5$

______________ 7. $x = 2y^2 + 3y + 10$

______________ 8. $3x^2 + y^2 = 9$

Graph the following inequalities. Sketch the conic section and shade the appropriate region.

9. $x^2 + y^2 < 16$

conic section:

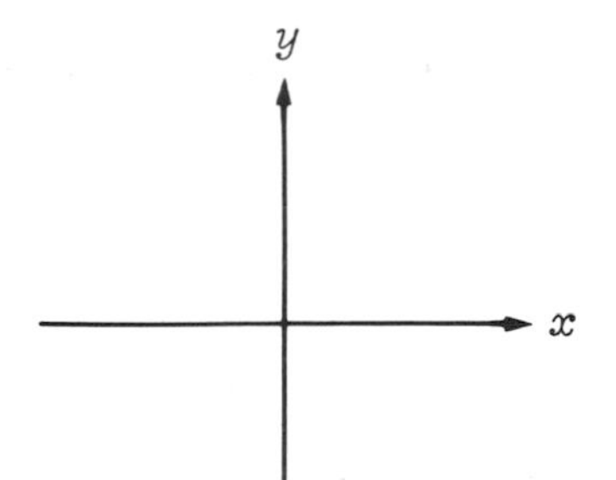

10. $x \leq y^2 + 3y$

conic section: parabola

$x \leq (y^2 + 3y + \frac{9}{4}) - \frac{9}{4}$

$x \leq (y + \frac{3}{2})^2 - \frac{9}{4}$

vertex: $(-2\frac{1}{4}, -1\frac{1}{2})$ opens to right

axis: $y = -1\frac{1}{2}$

x	y
0	0
0	-3
4	1
4	-4

Test: $(4, -2)$

$4 \leq (-2)^2 + 3(-2)$

$4 \leq 4 - 6$

$4 \leq -2$ False

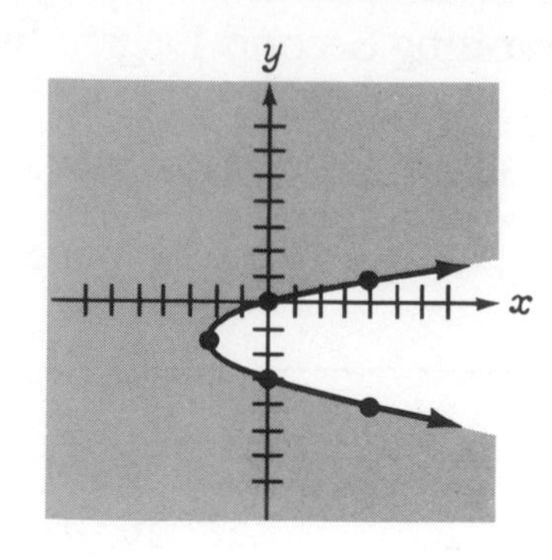

11. $\frac{3x^2}{27} + \frac{2y^2}{72} \geq 1$

conic section: ellipse

$\frac{x^2}{9} + \frac{y^2}{36} \geq 1$

Test: $(0,0)$

$\frac{0^2}{9} + \frac{0^2}{36} \geq 1$

$0 \geq 1$ False

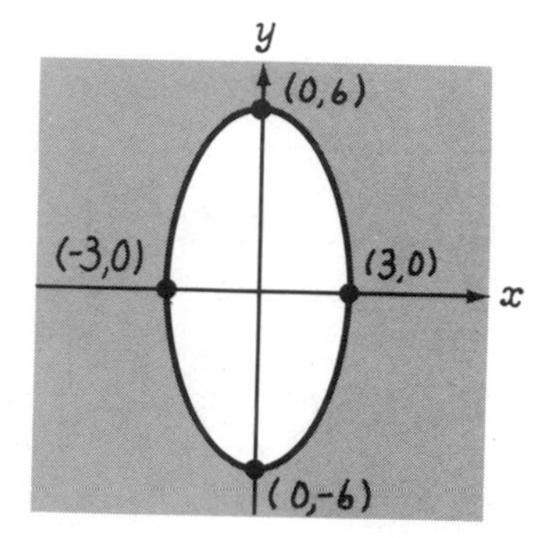

12. $\frac{(x - 2)^2}{4} - (y + 3)^2 > 1$

conic section: hyperbola

center: $(2, -3)$

vertices: $(4, -3)$ and $(0, -3)$

Test: $(0,0)$

$\frac{(0-2)^2}{4} - (0+3)^2 > 1$

$1 - 9 > 1$

$-8 > 1$ False

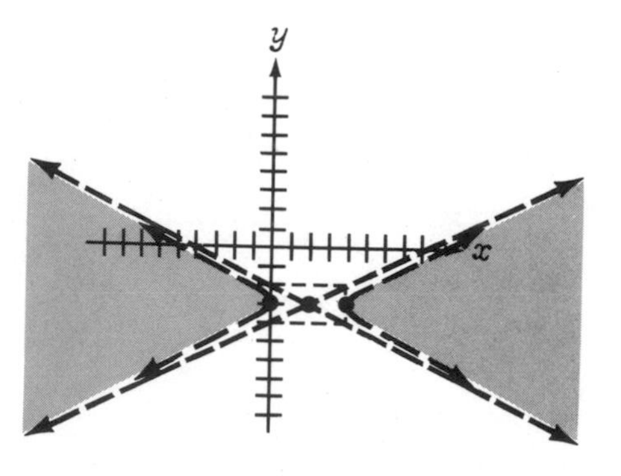

10. $x \leq y^2 + 3y$

conic section:

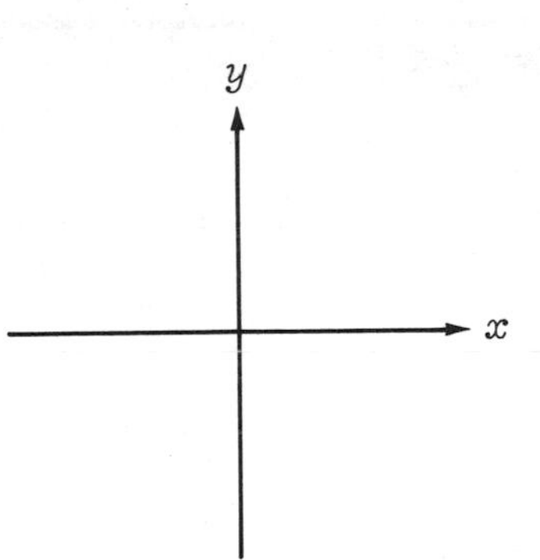

11. $\frac{3x^2}{27} + \frac{2y^2}{72} \geq 1$

conic section:

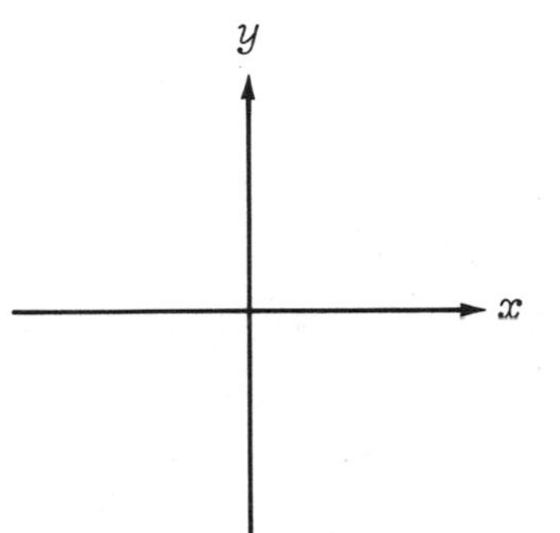

12. $\frac{(x - 2)^2}{4} - (y + 3)^2 > 1$

conic section:

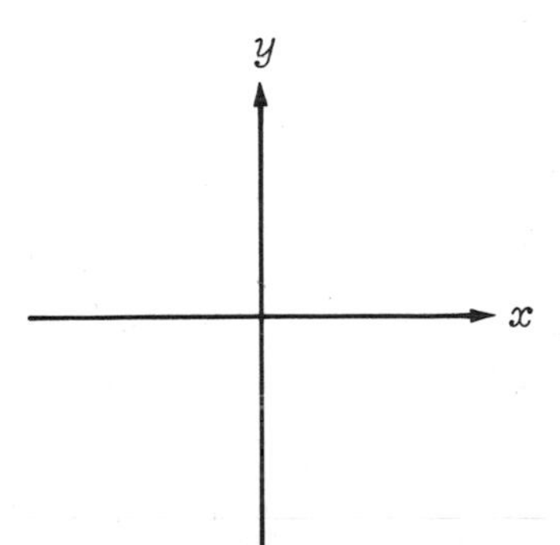

Chapter 8 Self-Test

Graph each equation or inequality.

1. $x = 2/3\ y^2 - 2y - 12$

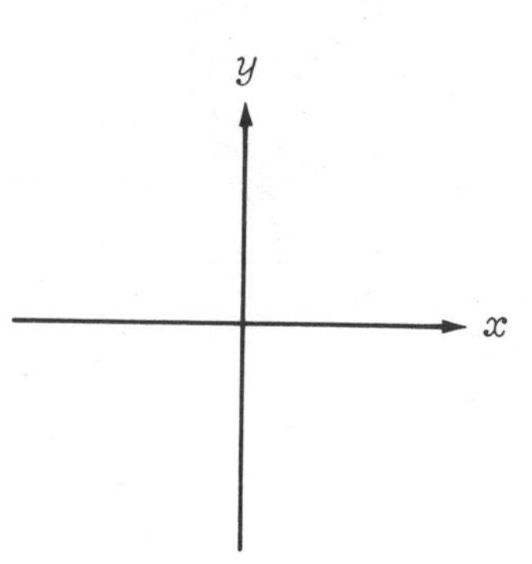

2. $x^2 + y^2 - 4x + y - 19/4 = 0$

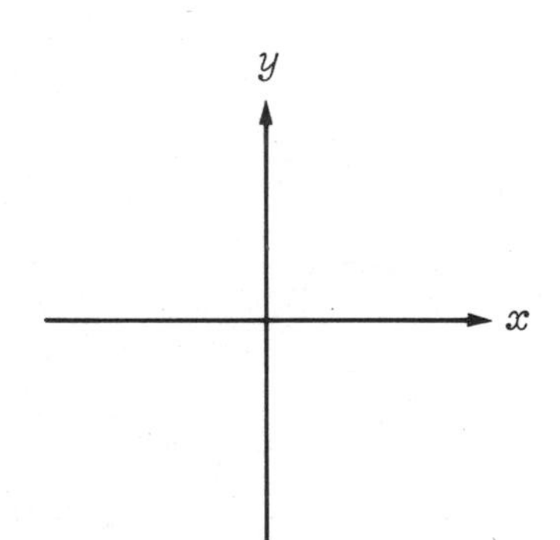

3. $4x^2 + 9y^2 = 36$

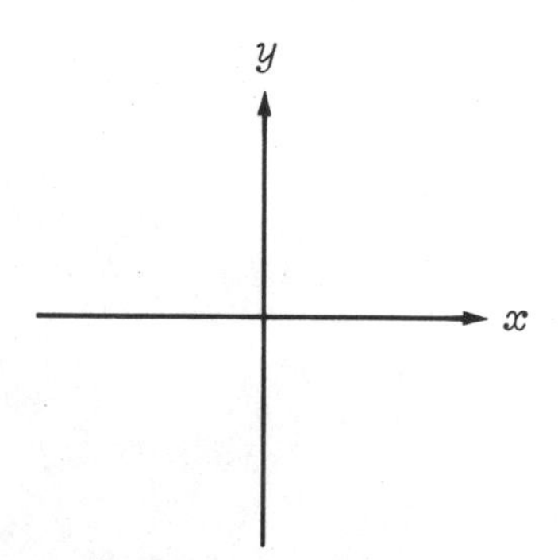

4. $12x^2 - 3y^2 = 48$

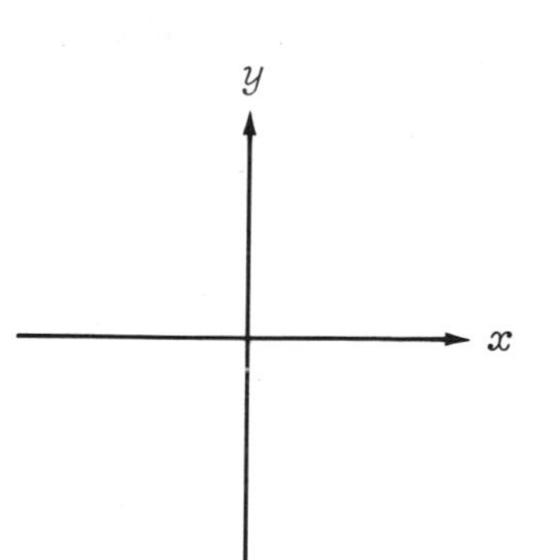

5. $x \geq 2y^2 + 7$

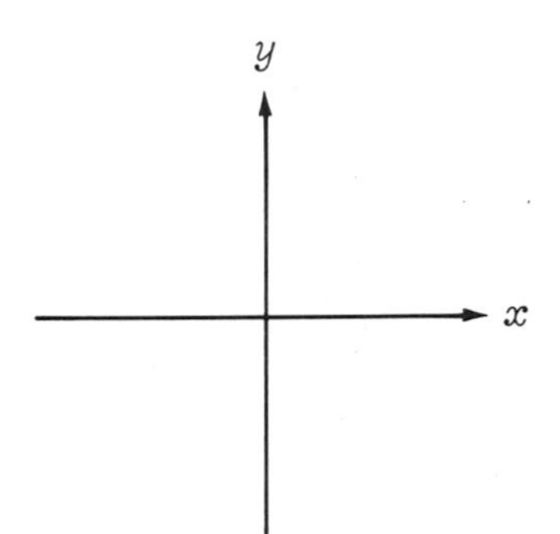

6. $x^2 + y^2 - 4x - 6y + 9 \leq 0$

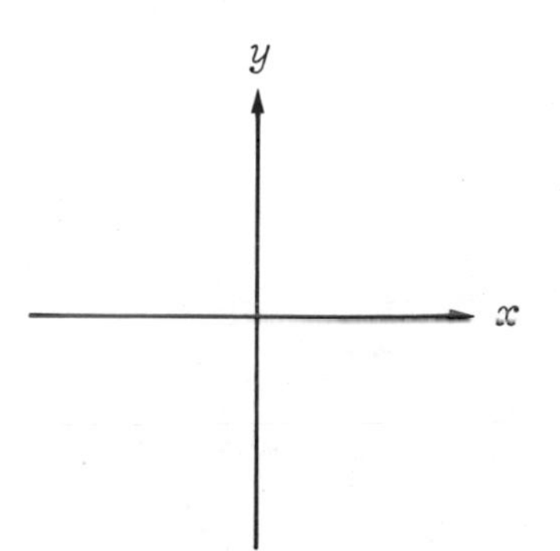

CHAPTER

9 Systems of Equations

9 · 1 Linear Systems in Two Variables

1. The solutions of a system of equations are the ordered pairs that satisfy every equation in the system.
2. If the graph of a linear system results in one point of intersection, then there is only 1 solution of the system, and the system is said to be independent. When solved algebraically, an independent system will result in only one value for x and only one value for y, which is one ordered pair (x, y).
3. If the graph of a linear system results in parallel lines, then there is no point of intersection. There is no solution of the system and the system is said to be inconsistent. When solved algebraically, an inconsistent system will result in a false statement of equality, and there will be no solution.
4. If the graph of a linear system results in lines that coincide (that is, the graphs of all the equations in the system are the same line),

C H A P T E R

9 Systems of Equations

9·1 Linear Systems in Two Variables

1. The ______________ of a system of equations are the ____________ __________ that satisfy every equation in the system.

2. If the graph of a linear system results in one point of intersection, then there is only _____ solution of the system, and the system is said to be _________________. When solved algebraically, an independent system will result in only one value for x and only one value for y, which is one ordered pair (x, y).

3. If the graph of a linear system results in parallel lines, then there is ________ point of intersection. There is no solution of the system and the system is said to be __________________. When solved algebraically, an inconsistent system will result in a false statement of equality, and there will be no solution.

4. If the graph of a linear system results in lines that coincide (that is, the graphs of all the equations in the system are the same line),

there is an infinite number of solutions of the system, and the system is said to be dependent. When solved algebraically, a dependent system will result in a true statement of equality, and there will be an infinite number of solutions.

5. $2x + y = 7$

$x - y = -1$

(a) Solve the system by the graphing method.

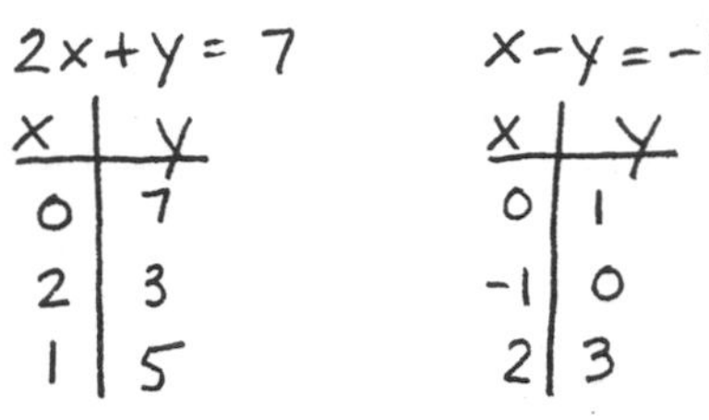

$2x+y=7$

x	y
0	7
2	3
1	5

$x-y=-1$

x	y
0	1
-1	0
2	3

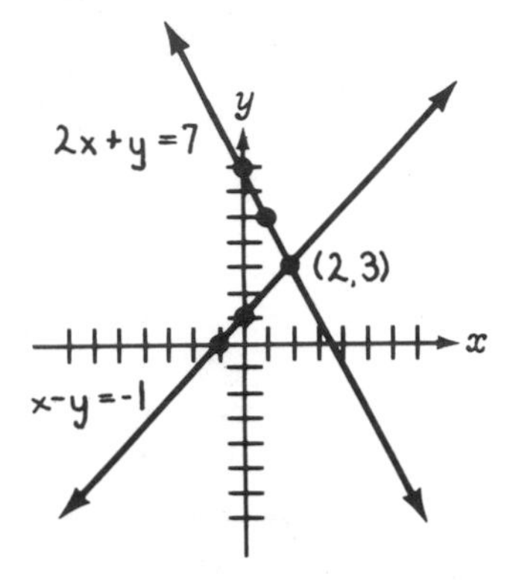

$(2,3)$

(point of intersection)

(b) Solve the system by the substitution method. Solve for y in $2x + y = 7$ as the first step.

$2x+y=7$

$y=-2x+7$

$x-y=-1$

$x-(-2x+7)=-1$

$3x-7=-1$

$3x=6$

$x=2$

$y=-2(2)+7$

$y=3$

$(2,3)$

(c) Solve the system by the elimination method. Solve by eliminating y as a first step.

$2x+y=7$

$x-y=-1$

$3x = 6$

$x = 2$

$2(2)+y=7$

$4+y=7$

$y=3$

$(2,3)$

there is an infinite number of solutions of the system, and the system is said to be ______________. When solved algebraically, a dependent system will result in a true statement of equality, and there will be an infinite number of solutions.

5. $2x + y = 7$

   $x - y = -1$

   (a) Solve the system by the graphing method.

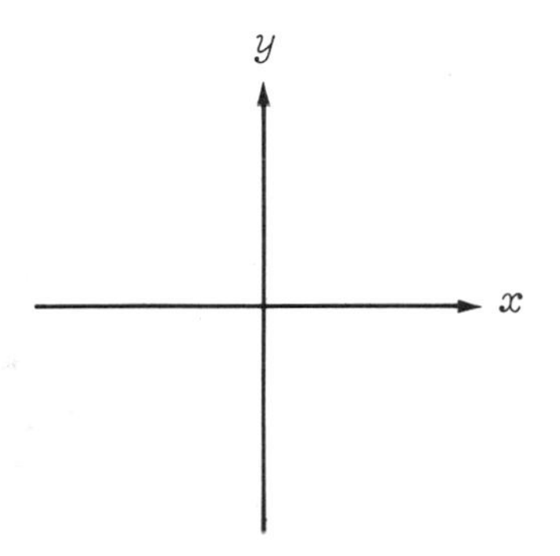

   (b) Solve the system by the substitution method. Solve for y in $2x + y = 7$ as the first step.

   (c) Solve the system by the elimination method. Solve by eliminating y as a first step.

(d) This system is independent. (dependent, inconsistent, or independent)

6.
33

$-5x - 2y = 6$

$2x - 5y = -2$

(a) Solve the system by the graphing method.

$-5x - 2y = 6$

x	y
0	-3
1	$-5\frac{1}{2}$
$-\frac{6}{5}$	0

$2x - 5y = -2$

x	y
0	$\frac{2}{5}$
-1	0

Note: the graphing method is not exact for problems such as this, where it is difficult to find the exact coordinates of the point of intersection.

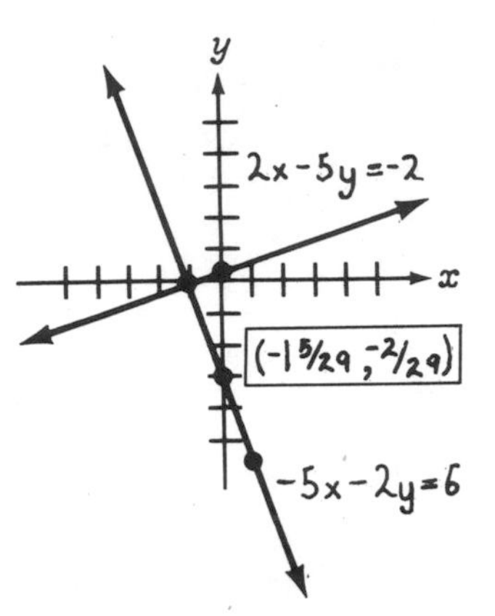

(b) Solve the system by the substitution method. Solve for x in $-5x - 2y = 6$ as the first step.

$-5x - 2y = 6$

$-5x = 2y + 6$

$x = -\frac{2}{5}y - \frac{6}{5}$

$\boxed{\left(-\frac{34}{29}, -\frac{2}{29}\right)}$

$2x - 5y = -2$

$2\left(-\frac{2}{5}y - \frac{6}{5}\right) - 5y = -2$

$-\frac{4}{5}y - \frac{12}{5} - 5y = -2$

$-4y - 12 - 25y = -10$

$-29y = 2 \quad y = \frac{-2}{29}$

$-5x - 2y = 6$

$-5x - 2\left(\frac{-2}{29}\right) = 6$

$-5x + \frac{4}{29} = 6$

$-145x + 4 = 174$

$-145x = 170$

$x = \frac{-170}{145} = \frac{-34}{29}$

(d) This system is ________________. (dependent, inconsistent, or independent)

6. **33**

$-5x - 2y = 6$

$2x - 5y = -2$

(a) Solve the system by the graphing method.

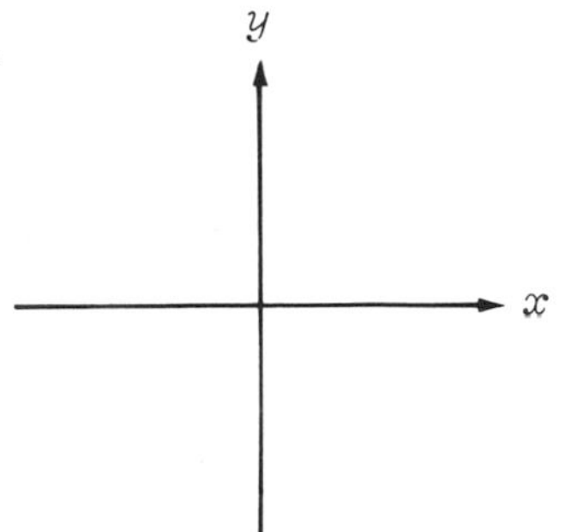

(b) Solve the system by the substitution method. Solve for x in $-5x - 2y = 6$ as the first step.

(c) Solve the system by the elimination method. Solve by eliminating x as the first step.

$$-5x - 2y = 6 \quad \Rightarrow \quad -10x - 4y = 12$$
$$2x - 5y = -2 \quad \Rightarrow \quad 10x - 25y = -10$$
$$-29y = 2$$
$$y = -\frac{2}{29}$$

$$-5x - 2\left(\frac{-2}{29}\right) = 6$$
$$-5x + \frac{4}{29} = 6$$
$$-5x = 5\frac{25}{29}$$
$$x = -1\frac{5}{29}$$

$$\boxed{\left(-1\frac{5}{29}, -\frac{2}{29}\right)}$$

(d) This system is independent. (dependent, inconsistent, or independent)

7. $1/2\ x + 2/3\ y = 6$

$1/4\ x + 1/3\ y = 3$

(a) Solve the system by the graphing method.

x	y
0	9
12	0

x	y
0	9
12	0

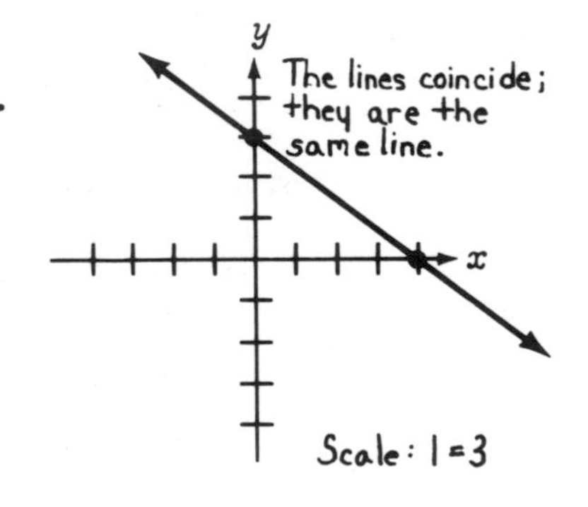

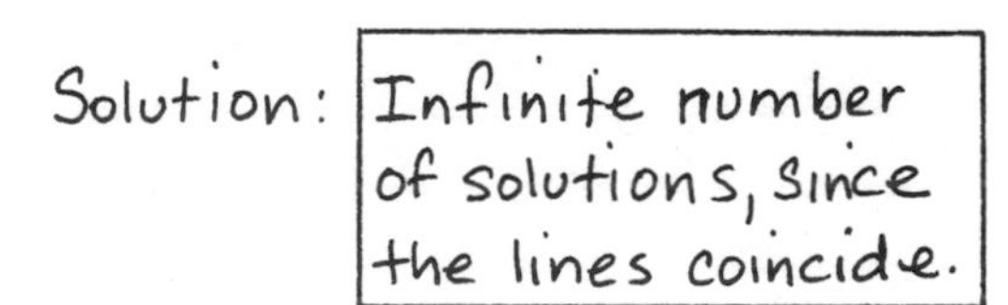

(b) Solve the system by the substitution method.

$$\frac{1}{2}x + \frac{2}{3}y = 6$$
$$\frac{1}{2}x = -\frac{2}{3}y + 6$$
$$x = -\frac{4}{3}y + 12$$

$$\frac{1}{4}x + \frac{1}{3}y = 3$$
$$\frac{1}{4}\left(-\frac{4}{3}y + 12\right) + \frac{1}{3}y = 3$$
$$-\frac{1}{3}y + 3 + \frac{1}{3}y = 3$$
$$3 = 3 \quad \text{True}$$

Infinite number of solutions.

(c) Solve the system by the elimination method. Solve by eliminating x as the first step.

(d) This system is 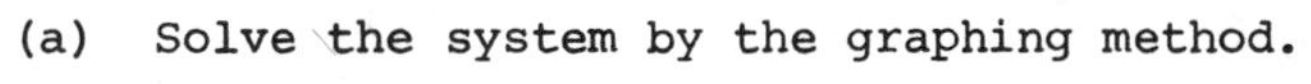 (dependent, inconsistent, or independent)

7. $1/2\ x + 2/3\ y = 6$

$1/4\ x + 1/3\ y = 3$

(a) Solve the system by the graphing method.

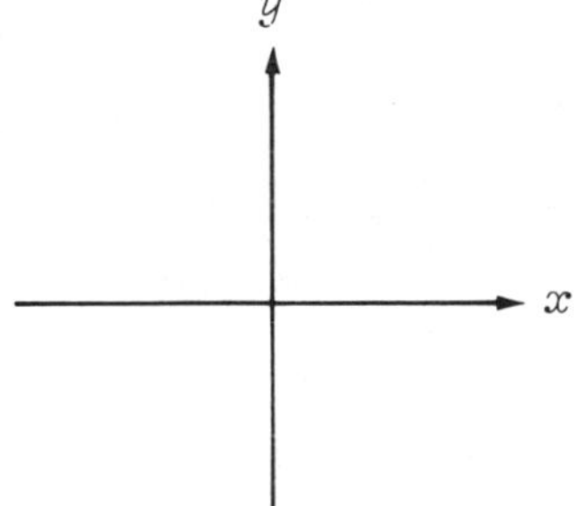

(b) Solve the system by the substitution method.

(c) Solve the system by the elimination method.

$$\frac{1}{2}x + \frac{2}{3}y = 6 \quad \Rightarrow \quad \frac{1}{2}x + \frac{2}{3}y = 6$$
$$\frac{1}{4}x + \frac{1}{3}y = 3 \quad\quad\quad -\frac{1}{2}x - \frac{2}{3}y = -6$$
$$0 = 0 \quad \text{True}$$

Infinite number of solutions.

(d) This system is dependent. (dependent, inconsistent, or independent)

8. $3x + y = 6$

$6x + 2y = 8$

(a) Solve the system by the graphing method.

$3x + y = 6$

x	y
0	6
2	0
1	3

$6x + 2y = 8$

x	y
0	4
$\frac{4}{3}$	0
2	-2

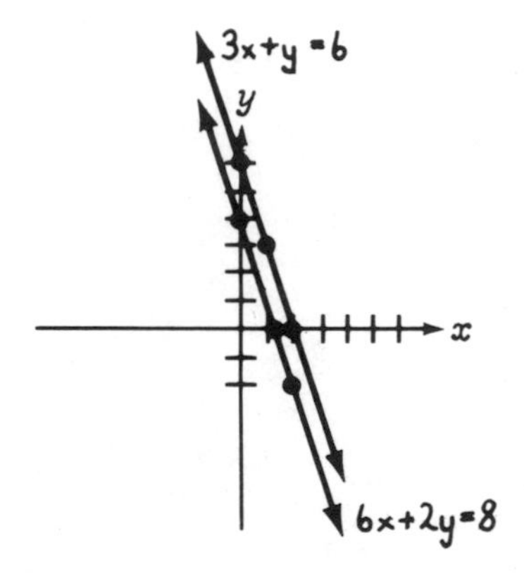

Parallel lines - no solution.

(b) Solve the system by the substitution method.

$3x + y = 6$
$y = -3x + 6$

$6x + 2y = 8$
$6x + 2(3x + 6) = 8$
$6x - 6x + 12 = 8$
$12 = 8$ False
$12 \neq 8$

No solution

(c) Solve the system by the elimination method.

(d) This system is ______________. (dependent, inconsistent, or independent)

8. $3x + y = 6$

$6x + 2y = 8$

(a) Solve the system by the graphing method.

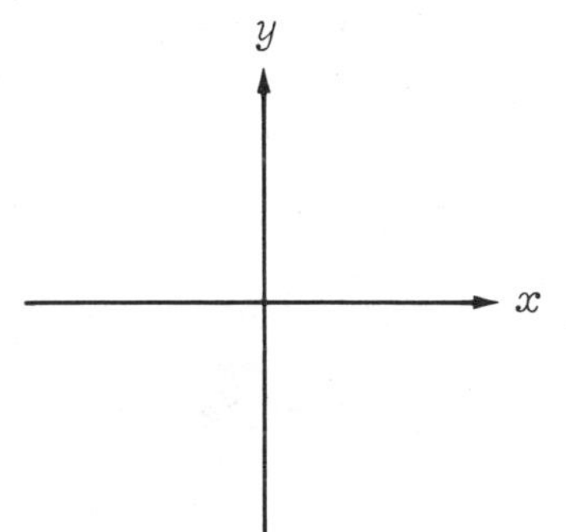

(b) Solve the system by the substitution method.

(c) Solve the system by the elimination method.

$3x+y=6 \Rightarrow 6x+2y=12$

$6x+2y=8 \Rightarrow 6x+2y=8$

$0=-4$ False

$0\neq -4$

No solution

(d) This system is inconsistent. (dependent, inconsistent, or independent)

9·2 Linear Systems in Three Variables

1. $2x + y - z = -4$

$x - y + 2z = 15$

$3x - 2y - z = 7$

Solve this system by eliminating x as the first step in the following pairs of equations.

$$\begin{cases} 2x + y - z = -4 \\ x - y + 2z = 15 \end{cases} \qquad \begin{cases} x - y + 2z = 15 \\ 3x - 2y - z = 7 \end{cases}$$

$$\begin{cases} 2x+y-z=-4 \\ -2x+2y-4z=-30 \end{cases} \qquad \begin{cases} -3x+3y-6z=-45 \\ 3x-2y-z=7 \end{cases}$$

$3y-5z=-34 \qquad y-7z=-38$

$3y-5z=-34 \Rightarrow 3y-5z=-34$

$y-7z=-38 \Rightarrow -3y+21z=114$

$16z=80$

$z=5$

$3y-5z=-34$

$3y-5(5)=-34$

$3y-25=-34$

$3y=-9 \quad y=-3$

$2x+y-z=-4$

$2x+(-3)-(5)=-4$

$2x-8=-4$

$2x=4 \quad x=2$

$(2,-3,5)$

This system is independent. (independent, dependent, or inconsistent)

(c) Solve the system by the elimination method.

(d) This system is ________________. (dependent, inconsistent, or independent)

9·2 Linear Systems in Three Variables

1. $2x + y - z = -4$

   $x - y + 2z = 15$

   $3x - 2y - z = 7$

   Solve this system by eliminating x as the first step in the following pairs of equations.

   $2x + y - z = -4$ $x - y + 2z = 15$

   $x - y + 2z = 15$ $3x - 2y - z = 7$

   This system is ________________. (independent, dependent, or inconsistent)

2. $1/2\ x - 1/3\ y + 1/4\ z = 12$

$2x - 4/3\ y + z = 48$

$-3/2\ x + y - 3/4\ z = -36$

Solve this system by eliminating z as the first step:

$$\begin{cases} 1/2\ x - 1/3\ y + 1/4\ z = 12 \\ 2x - 4/3\ y + z = 48 \end{cases} \qquad \begin{cases} 1/2\ x - 1/3\ y + 1/4\ z = 12 \\ -3/2\ x + y - 3/4\ z = -36 \end{cases}$$

$$-2x + \frac{4}{3}y - z = -48 \qquad \frac{3}{2}x - y + \frac{3}{4}z = 36$$

$$2x - \frac{4}{3}y - z = 48 \qquad -\frac{3}{2}x + y - \frac{3}{4}z = -36$$

$0 = 0$ True $\qquad 0 = 0$ True

Infinite number of solutions.

This system is dependent. (independent, dependent, or inconsistent)

9.3 Applications

1. / 3 The sum of two numbers is 6, and their difference is 10. What are the numbers? Solve by the graphing method.

$x + y = 6$

x	y
2	4
4	2

$x - y = 10$

x	y
5	-5
4	-6

8 and -2

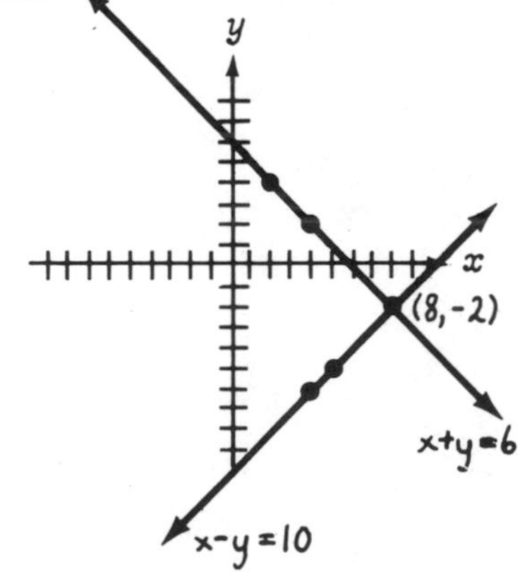

2. $1/2\,x - 1/3\,y + 1/4\,z = 12$

$2x - 4/3\,y + z = 48$

$-3/2\,x + y - 3/4\,z = -36$

Solve this system by eliminating z as the first step:

$1/2\,x - 1/3\,y + 1/4\,z = 12$ $\qquad$ $1/2\,x - 1/3\,y + 1/4\,z = 12$

$2x - 4/3\,y + z = 48$ $\qquad$ $-3/2\,x + y - 3/4\,z = -36$

This system is ______________. (independent, dependent, or inconsistent)

9.3 Applications

1. The sum of two numbers is 6, and their difference is 10. What are the numbers? Solve by the graphing method.

y

x

2.
7

Maggie has \$2.80 in nickels and dimes. She has a total of 40 coins. How many nickels and how many dimes does Maggie have? Solve by the substitution method.

$x = \#\text{nickels}$ $\quad y = \#\text{ dimes}$

$x + y = 40 \Rightarrow y = -x + 40$

$.05x + .10y = 2.80 \Rightarrow 5x + 10y = 280$

$5x + 10(-x + 40) = 280$

$5x - 10x + 400 = 280$

$-5x = -120$

$x = 24$ nickels

$x + y = 40$

$24 + y = 40$

$y = 16$ dimes

24 nickels 16 dimes

3.
17

At Athletic World, Coach Reeves buys 3 basketballs and 2 footballs for \$60, and Coach Johnson buys 2 basketballs and 5 footballs for \$73. What is the cost of each football and basketball? Solve using Cramer's Rule.

x = cost of one football

y = cost of one basketball

$2x + 3y = 60 \longrightarrow -4x - 6y = -120$

$5x + 2y = 73 \longrightarrow 15x + 6y = 219$

$11x = 99$

$x = \$9$

$2x + 3y = 60$

$2(9) + 3y = 60$

$3y = 42$

$y = \$14$

Cost of a football = \$9 Cost of a basketball = \$14

2. | 7

Maggie has $2.80 in nickels and dimes. She has a total of 40 coins. How many nickels and how many dimes does Maggie have? Solve by the substitution method.

3. | 17

At Athletic World, Coach Reeves buys 3 basketballs and 2 footballs for $60, and Coach Johnson buys 2 basketballs and 5 footballs for $73. What is the cost of each football and basketball? Solve using Cramer's Rule.

4. **23** Patti invested \$7000 at her local savings-and-loan. She invested part of her money in an account paying 9% simple interest; she invested the rest of her money in an account paying 11% simple interest. After one year her interest income was \$750. How much did Patti invest in each account? Solve by elimination.

x = amount @ 9%
y = amount @ 11%

$$x + y = 7000 \Rightarrow 9x + 9y = 63000$$
$$.09x + .11y = 750 \Rightarrow {}^{-}9x + {}^{-}11y = {}^{-}75000$$
$$-2y = -12000$$
$$y = \$6000$$
$$x + y = 7000$$
$$x = \$1000$$

\$1000 @ 9%
\$6000 @ 11%

9.4 Nonlinear Systems of Equations

Determine the solutions of the following systems of equations. Solve by the method designated in each problem.

1. $x^2 + y^2 = 9$ Use the substitution method.

$2x - y = 3$

$$-y = -2x + 3$$
$$y = 2x - 3$$

$$x^2 + y^2 = 9$$
$$x^2 + (2x-3)^2 = 9$$
$$x^2 + 4x^2 - 12x + 9 = 9$$
$$5x^2 - 12x = 0$$
$$x(5x - 12) = 0$$
$x = 0$ or $5x - 12 = 0$ $\quad 5x = 12$

or $x = \frac{12}{5}$

If $x = 0$: $x^2 + y^2 = 9$
$$0^2 + y^2 = 9$$
$$y^2 = 9 \quad y = \pm 3$$
Extraneous ~~$(0,3)$~~ or $(0,-3)$

If $x = \frac{12}{5}$: $x^2 + y^2 = 9$
$$\frac{144}{25} + y^2 = \frac{225}{25}$$
$$y^2 = \frac{81}{25} \quad y = \pm\frac{9}{5}$$
$\left(\frac{12}{5}, \frac{9}{5}\right)$ or ~~$\left(\frac{12}{5}, -\frac{9}{5}\right)$~~
Extraneous

$(0,-3)$ or $\left(\frac{12}{5}, \frac{9}{5}\right)$

4. 23 Patti invested $7000 at her local savings-and-loan. She invested part of her money in an account paying 9% simple interest; she invested the rest of her money in an account paying 11% simple interest. After one year her interest income was $750. How much did Patti invest in each account? Solve by elimination.

9.4 Nonlinear Systems of Equations

Determine the solutions of the following systems of equations. Solve by the method designated in each problem.

1. $x^2 + y^2 = 9$ Use the substitution method.

   $2x - y = 3$

2\. 25

$$\begin{cases} 5x^2 + 2y^2 = 7 \\ x^2 + y^2 = 2 \end{cases}$$

Use the elimination method.

$$5x^2 + 2y^2 - 7$$
$$-5x^2 - 5y^2 = -10$$
$$-3y^2 = -3$$
$$y^2 = 1$$
$$y = \pm 1$$

If $y = 1$: $x^2 + (1)^2 = 2$
$$x^2 = 1$$
$$x = \pm 1$$
$(1,1)$ OR $(-1,1)$

If $y = -1$: $x^2 + (-1)^2 = 2$
$$x^2 = 1$$
$$x = \pm 1$$
$(1,-1)$ OR $(-1,-1)$

$$\boxed{(1,1),(-1,1),(1,-1),(-1,-1)}$$

3\. 33

$$\begin{cases} x^2 - 4y^2 = -16 \\ 2x - y^2 = 0 \end{cases}$$

Use the elimination method.

$$x^2 - 4y^2 = -16$$
$$-8x + 4y^2 = 0$$
$$x^2 - 8x = -16$$
$$x^2 - 8x + 16 = 0$$
$$(x-4)^2 = 0$$
$$x = 4$$

$$2x - y^2 = 0$$
$$2(4) - y^2 = 0$$
$$y^2 = 8$$
$$y = \pm 2\sqrt{2}$$

$$\boxed{(4, 2\sqrt{2}),(4, -2\sqrt{2})}$$

2.
25

$$5x^2 + 2y^2 = 7$$
$$x^2 + y^2 = 2$$

Use the elimination method.

3.
33

$$x^2 - 4y^2 = -16$$
$$2x - y^2 = 0$$

Use the elimination method.

4. | 43

The perimeter of a rectangle is 32 feet. The rectangle has an area of 48 square feet. What are the dimensions of the rectangle? Use the substitution method.

$x = \text{length} \quad y = \text{width}$

$2x + 2y = 32 \quad \Rightarrow \quad 2x = 32 - 2y$

$xy = 48 \qquad\qquad x = 16 - y$

$(16 - y)y = 48$

$16y - y^2 = 48$

$y^2 - 16y + 48 = 0$

$(y - 12)(y - 4) = 0$

$y = 12$ OR $y = 4$

$xy = 48 \qquad xy = 48$

$x = 4 \qquad x = 12$

4 ft. × 12 ft.

4. / 43 The perimeter of a rectangle is 32 feet. The rectangle has an area of 48 square feet. What are the dimensions of the rectangle? Use the substitution method.

Chapter 9 Self-Test

Solve each system of equations.

1. $x/3 - y/2 = 1$

   $4x - 6y = 12$

2. $2x - 5y = 3$

   $7x - 2y = 4$

3. $2x - 5y = 6$

   $16x - 40y = 10$

4. $x + 3y = 5$

   $2x - 3y + 2z = 0$

   $5x - 3z = 21$

5. $x^2 + y^2 = 13$

   $x - 2y = 4$

6. $4x^2 + y^2 = 36$

   $y = -x^2 + 1$

Solve the following systems.

7. $$\begin{aligned} 3x - 2y &= 11 \\ -2x + 7y &= 4 \end{aligned}$$

8. $$\begin{aligned} 2x + 3y + z &= -7 \\ -x + 2y - 5z &= -24 \\ 3x - 2y - 3z &= 2 \end{aligned}$$

9. Sam had $10,000 to invest. He invested part of his money in an account paying 7% simple interest, and the balance in an account paying 4%. After one year, his interest income was $520. How much money did he invest in each account?

C H A P T E R

10 Exponential and Logarithmic Functions

10 · 1 Inverse Functions

1. If $f(x) = 2x + 5$, find the inverse of $f(x)$.

$y = 2x + 5$ -------> $x = 2y + 5$ (1) Interchange x and y.

$x - 5 = 2y$ (2) Solve for y.

$\frac{x-5}{2} = y$

$\frac{x-5}{2} = f^{-1}(x)$

2. Given $f(x) = \sqrt{x - 4}$. The domain of $f(x)$ is $\{x | x \geq 4\}$ and the range of $f(x)$ is $\{y | y \geq 0\}$.

Find the inverse of $f(x)$.

$y = \sqrt{x - 4}$ -------> $x = \sqrt{y - 4}$

$x^2 = y - 4$

$x^2 + 4 = y$

This inverse will have a domain of $\{x | x \geq 0\}$ and a range of $\{y | y \geq 4\}$. Is this inverse a function? yes. Why? For every x, there is only one y (note restriction on x).

CHAPTER

10 Exponential and Logarithmic Functions

10 · 1 Inverse Functions

1. If $f(x) = 2x + 5$, find the inverse of $f(x)$.

   $y = 2x + 5$ -------> $x = 2y + 5$ (1) Interchange x and y.

   _______ $= 2y$ (2) Solve for y.

   _______ $= y$

   _______ $= f^{-1}(x)$

2. Given $f(x) = \sqrt{x - 4}$. The domain of $f(x)$ is _______________

   and the range of $f(x)$ is _______________.

   Find the inverse of $f(x)$.

   $y = \sqrt{x - 4}$ -------> $x = \sqrt{y - 4}$

   _______ $= y - 4$

   _______ $= y$

   This inverse will have a domain of _____________ and a range of

   _____________. Is this inverse a function? ______. Why? _________

   ____

3. Given $f(x) = |x|$. The domain of $f(x)$ is $\{x | x \in R\}$ and the range of $f(x)$ is $\{y | y \geq 0\}$.

Find the inverse of $f(x)$.

$y = |x|$ -------> $x = |y|$

$x = y$ or $-x = y$

This inverse will have a domain of $\{x | x \geq 0\}$ and a range of $\{y | y \in R\}$. Is this inverse a function? No

Why? For every $x > 0$, there are two values of y

Is $f(x)$ a one-to-one function? No Why? f(x) can be one-to-one only if $f^{-1}(x)$ is a function.

Given each of the following functions, find the inverse. If the inverse is a function, write it in the form $y = f^{-1}(x)$. Graph both equations on the same coordinate system.

4. $f(x) = |x|$

$y = |x|$

x	y
0	0
1	1
-1	1

$x = |y|$

x	y
0	0
1	1
1	-1

Not a function; does not pass the vertical-line test.

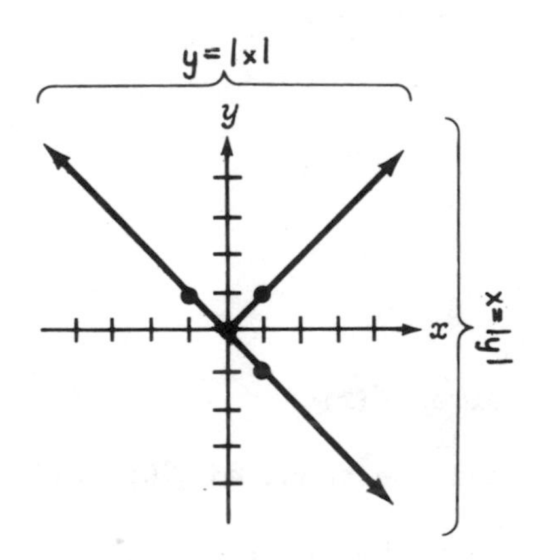

3. Given $f(x) = |x|$. The domain of $f(x)$ is ____________ and the range of $f(x)$ is ____________.

Find the inverse of $f(x)$.

$y = |x|$ -------> $x = |y|$

_____ $= y$ or _____ $= y$

This inverse will have a domain of ____________ and a range of ____________. Is this inverse a function? ________

Why? ____

Is $f(x)$ a one-to-one function? _____ Why? ______________________

____

Given each of the following functions, find the inverse. If the inverse is a function, write it in the form $y = f^{-1}(x)$. Graph both equations on the same coordinate system.

4. $f(x) = |x|$

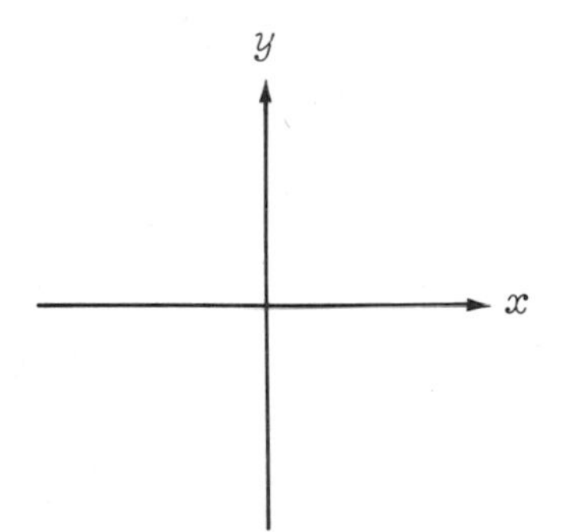

5. | 7

$f(x) = (2x - 5)/3$

$y = \frac{2x-5}{3}$ $\quad x = \frac{2y-5}{3}$ $\quad \frac{3x+5}{2} = y$

x	y
0	$-\frac{5}{3}$
1	-1
$\frac{4}{3}$	$-\frac{7}{9}$

x	y
$\frac{1}{3}$	3
-1	1
$-\frac{5}{3}$	0

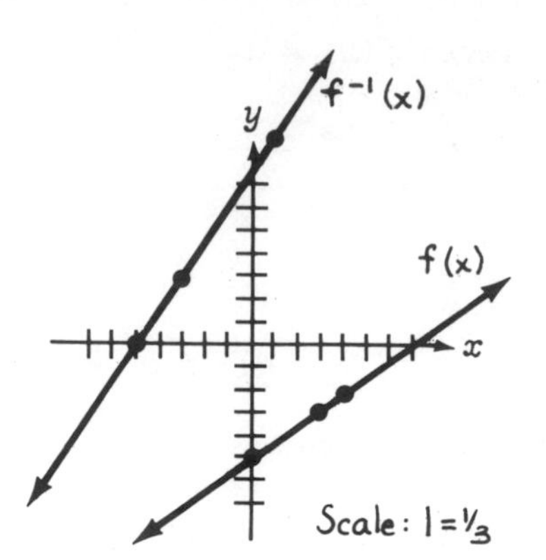

6. $f(x) = 9 - x^2$

$y = 9 - x^2$ $\quad x = 9 - y^2$

x	y
±2	5
0	9
±3	0

x	y
0	±3
5	±2
9	0

Not a function

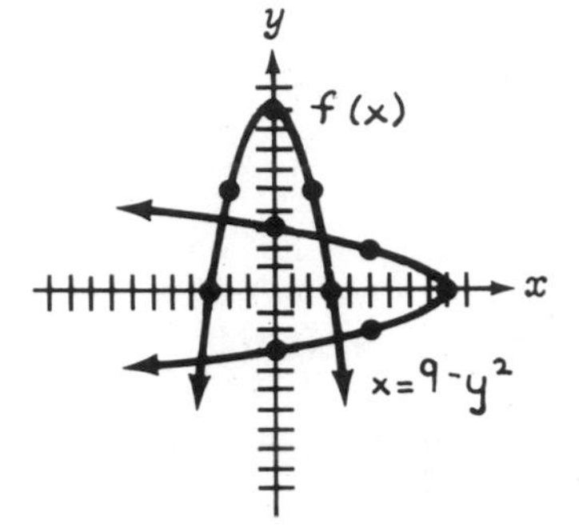

7. $f(x) = \sqrt{x + 2}$

$y = \sqrt{x+2}$ $\quad x = \sqrt{y+2}$

$x \geq -2$
$y \geq 0$

x	y
2	2
0	$\sqrt{2}$
-1	1
-2	0
7	3

x	y
0	-2
1	-1
2	2
$\sqrt{2}$	0
3	7

$x \geq 0$
$y \geq -2$

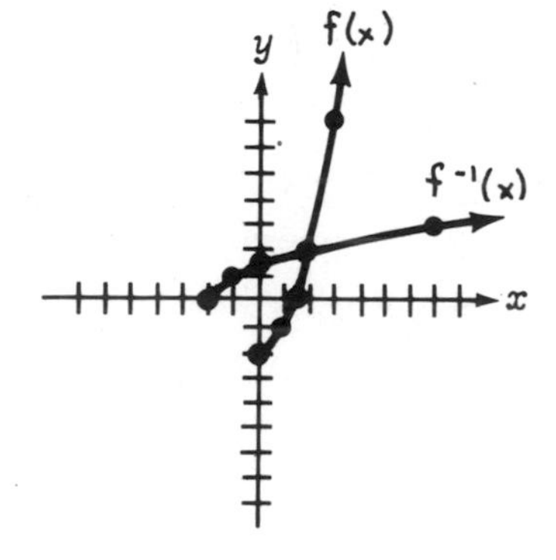

5. / 7 $f(x) = (2x - 5)/3$

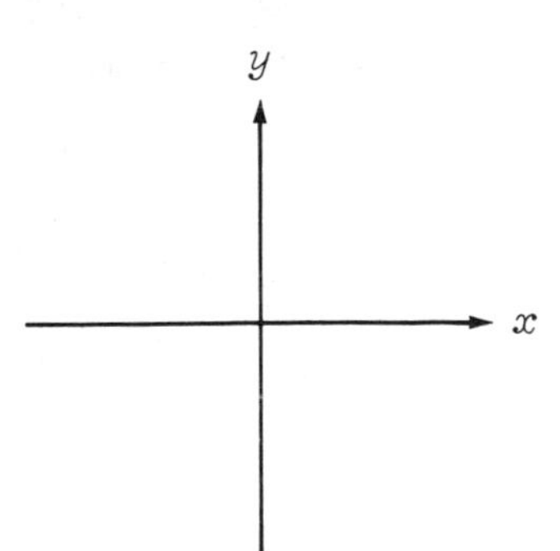

6. $f(x) = 9 - x^2$

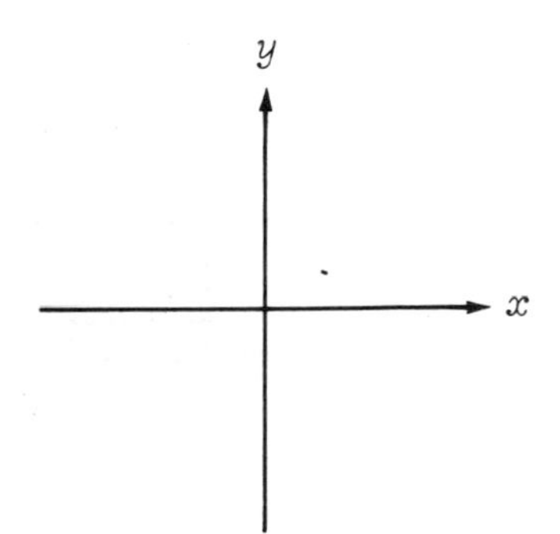

7. $f(x) = \sqrt{x + 2}$

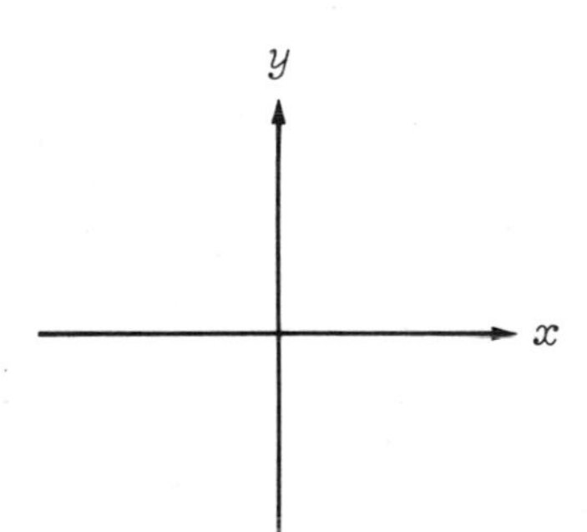

8. $f(x) = 2x + 1$ $g(x) = x^2 - 3$

$f(1) = 2(\underline{1}) + 1$

$f(a) = 2(\underline{a}) + 1$

$f(b^2) = 2(\underline{b^2}) + 1$

$f(1/2) = 2(\underline{\tfrac{1}{2}}) + 1$

$f(x^2 - 3) = 2(\underline{x^2-3}) + 1$

Therefore $f[\underline{g(x)}] = f(\underline{x^2 - 3}) = \underline{2(x^2-3)+1} = (f \circ g)(x)$

Given $f(x) = 3x - 1$ and $g(x) = x^2 + 1$ and $h(x) = (x + 1)/2$, find each of the following composite functions.

9. $(f \circ g)(x) = f(x^2+1)$
$= 3(x^2+1)-1$
$= 3x^2+3-1$
$= \boxed{3x^2+2}$

10. $(f \circ h)(x) = f\left(\frac{x+1}{2}\right)$
$= 3\left(\frac{x+1}{2}\right)-1$
$= \frac{3x+3}{2} - \frac{2}{2}$
$= \boxed{\frac{3x+1}{2}}$

11. $(f \circ g)(-2) = f((-2)^2+1)$
$= f(5)$
$f(5) = 3(5)-1 = \boxed{14}$

12. $(f \circ h)(4) =$
$f[h(4)] = f[\frac{4+1}{2}]$
$= f(\frac{5}{2})$
$f(\frac{5}{2}) = 3(\frac{5}{2}) - 1$
$= \frac{15}{2} - \frac{2}{2}$
$= \boxed{\frac{13}{2}}$

8. $f(x) = 2x + 1$ $\quad g(x) = x^2 - 3$

$f(1) = 2(\_\_\_) + 1$

$f(a) = 2(\_\_\_) + 1$

$f(b^2) = 2(\_\_\_) + 1$

$f(1/2) = 2(\_\_\_) + 1$

$f(x^2 - 3) = 2(\_\_\_\_\_\_\_\_) + 1$

Therefore $f[g(x)] = f(x^2 - 3) =$ ________________ $= (f \circ g)(x)$

Given $f(x) = 3x - 1$ and $g(x) = x^2 + 1$ and $h(x) = (x + 1)/2$, find each of the following composite functions.

9. $(f \circ g)(x)$

10. $(f \circ h)(x)$

11. $(f \circ g)(-2)$

12. $(f \circ h)(4)$

13. $(h \circ g)(x) = h[x^2+1]$

$= \frac{(x^2+1)+1}{2}$

$= \boxed{\frac{x^2+2}{2}}$

14. $(h \circ g)(0) =$

$h[g(0)] = h[0^2+1]$

$= h(1)$

$h(1) = \frac{1+1}{2} = \boxed{1}$

10 · 2 Exponential Functions and Equations

1. An exponential function is one in which the variable appears as an exponent.

2. If $f(x) = a^x$, then:

   (a) if $a > 1$, the graph rises from left to right.

   (b) if $a = 1$, the graph is the horizontal line $y = 1$.

   (c) if $0 < a < 1$, the graph rises from right to left.

   (d) Sketch the graph of each possibility listed above.

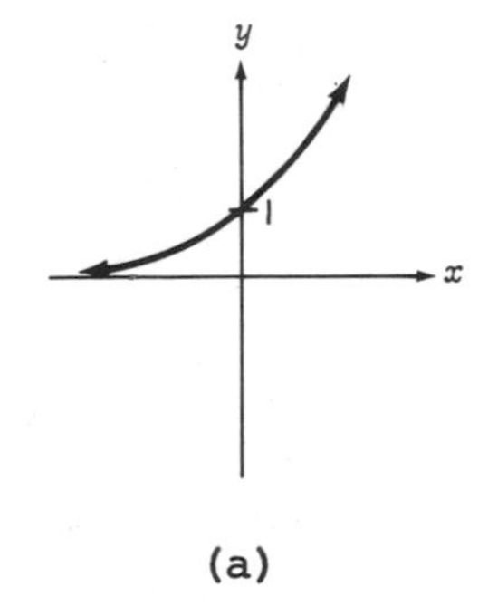

(a)

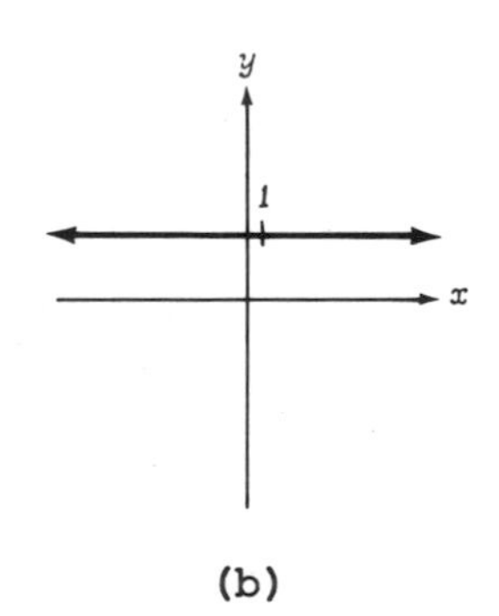

(b)

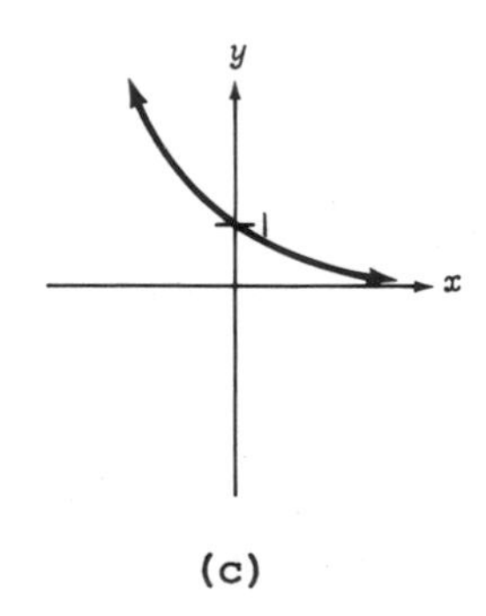

(c)

Sketch the graphs of the following functions.

3. [7] $f(x) = 3 + 2^x$

x	y
0	4
-1	3.5
1	5

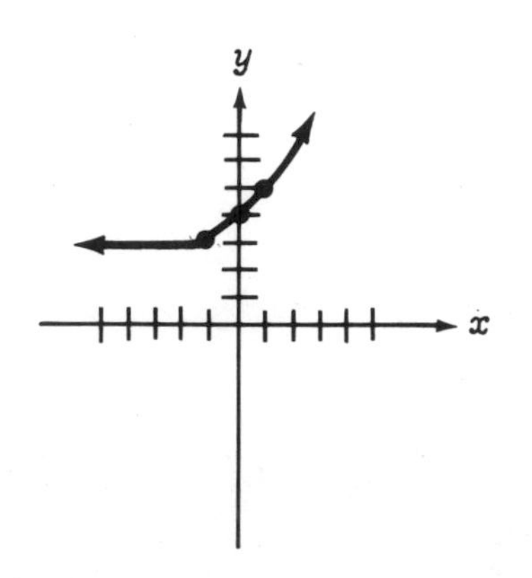

13. $(h \circ g)(x)$

14. $(h \circ g)(0)$

10·2 Exponential Functions and Equations

1. An exponential function is one in which the variable appears as an ____________.

2. If $f(x) = a^x$, then:

   (a) if $a > 1$, the graph rises from ________ to __________.

   (b) if $a = 1$, the graph is the ________________ line ______________.

   (c) if $0 < a < 1$, the graph rises from ___________ to __________.

   (d) Sketch the graph of each possibility listed above.

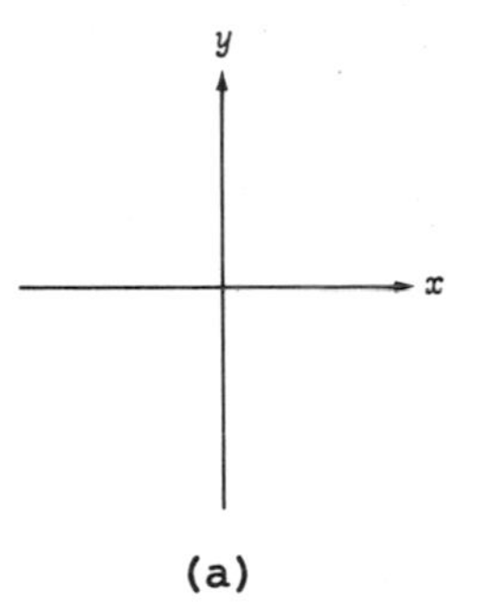

(a)

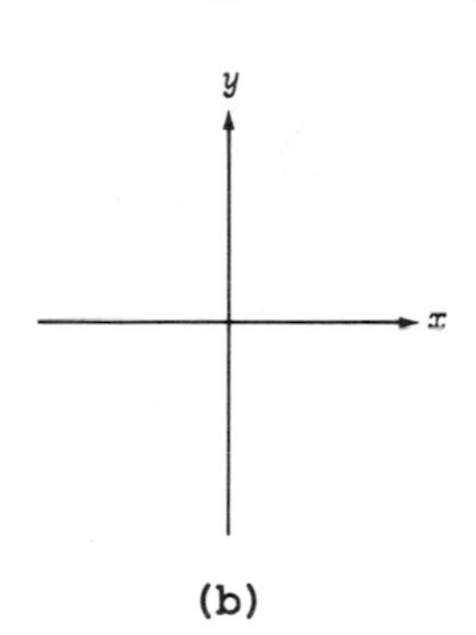

(b)

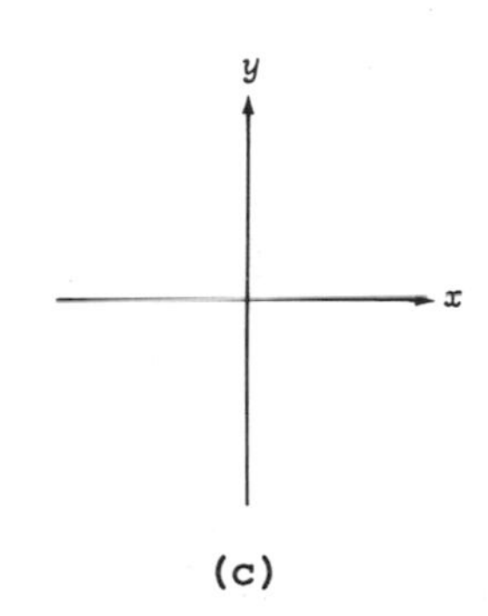

(c)

Sketch the graphs of the following functions.

3. 7 $f(x) = 3 + 2^x$

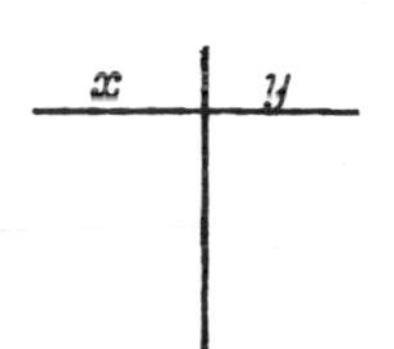

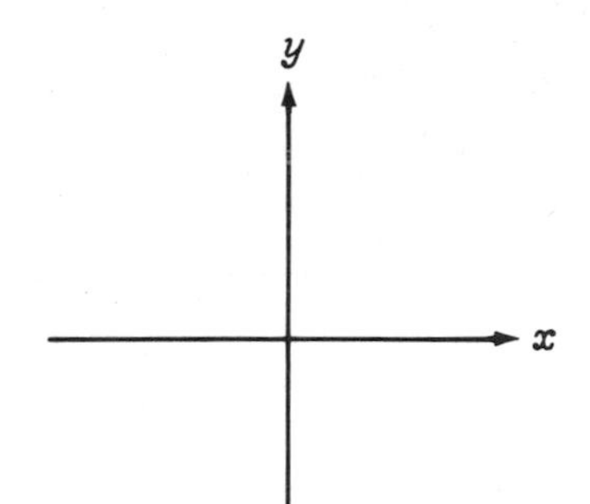

4. $f(x) = 2^{(x + 1)}$

x	y
-1	1
0	2
1	4
-2	$\frac{1}{2}$

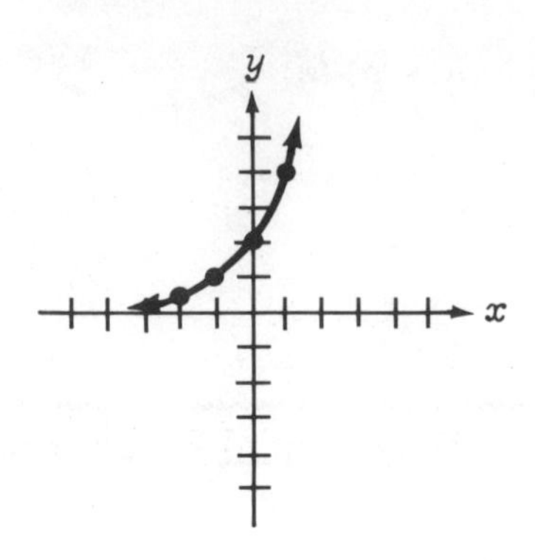

5. $f(x) = 2^{(-x/2)}$

x	y
-2	2
0	1
2	$\frac{1}{2}$
4	$\frac{1}{4}$
-4	4

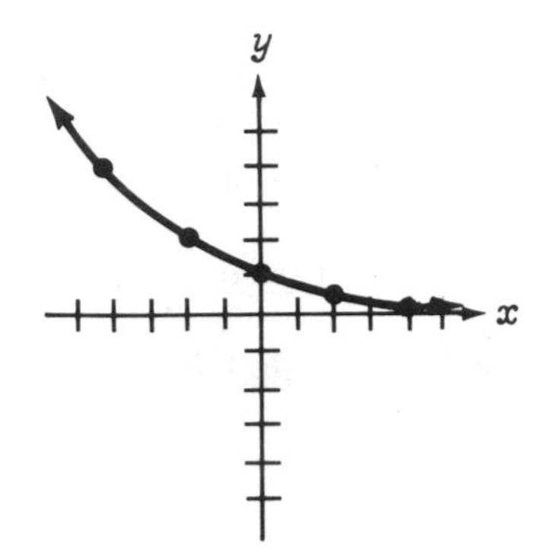

Solve the following equations. Remember that if $a^x = b^x$, then $a = b$.

6. | 28 | $81 = a^4$

$3^4 = a^4$

$\boxed{3 = a}$

7. $9 = a^{(-2/5)}$

$\left(\frac{1}{3}\right)^{-2} = \left(a^{\frac{1}{5}}\right)^{-2}$

$\left(\frac{1}{3}\right)^{5} = \left(a^{\frac{1}{5}}\right)^{5}$

$\boxed{\frac{1}{243} = a}$

4. $f(x) = 2^{(x + 1)}$

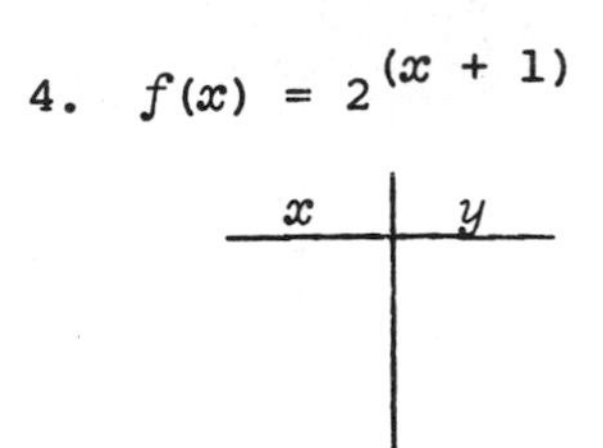

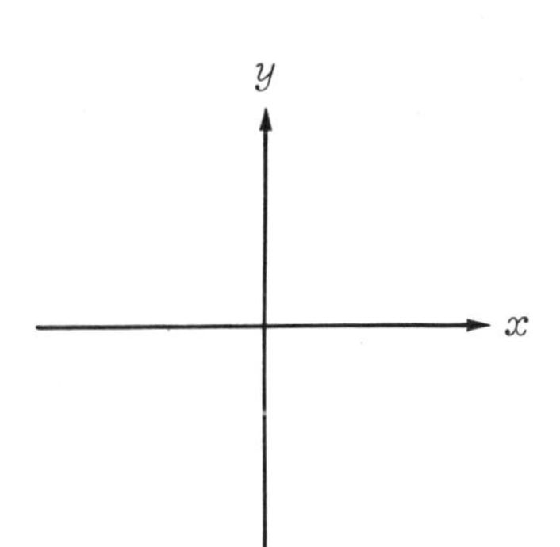

5. $f(x) = 2^{(-x/2)}$

x	y

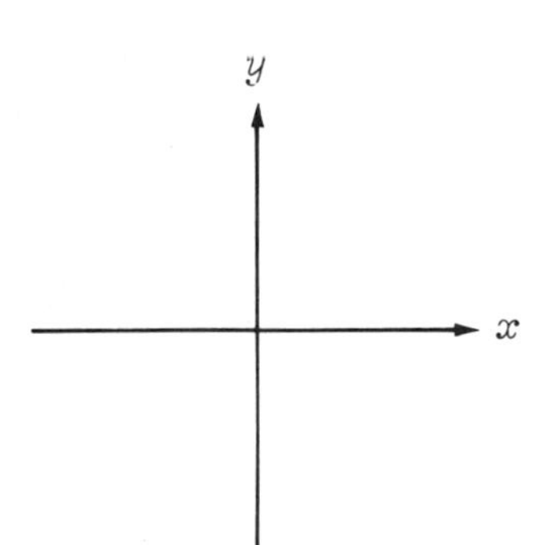

Solve the following equations. Remember that if $a^x = b^x$, then $a = b$.

6. 28 $81 = a^4$

7. $9 = a^{(-2/5)}$

8. $27 = 3^x$

$3^3 = 3^x$

$\boxed{x = 3}$

9. $32 = 2^{(2x + 1)}$

$2^5 = 2^{(2x+1)}$

$5 = 2x + 1$

$4 = 2x$

$\boxed{2 = x}$

10 · 3 Logarithmic Functions

1. Given $y = 3^x$, the inverse of $y = 3^x$ is $\underline{x = 3^y}$. Since it is not convenient to solve for y by conventional methods, the logarithmic function $y = \log_a x$ is defined. In this case, $x = 3^y$ could be written $\underline{y = \log_3 x}$.

2. If the logarithmic function $y = \log_a x$ is written in the form $y = \log_e x$, this is known as the natural logarithmic function and is written $\underline{y = \ln(x)}$.

Rewrite the following in logarithmic form.

3. $y = 2^e$ $\boxed{\log_2 y = e}$

4. $N = .035^{(x + 1)}$ $\boxed{\log_{.035} N = x+1}$

5. $K = e^{1.5}$ $\boxed{\log_e k = 1.5 \text{ OR } \ln k = 1.5}$

Rewrite the following in exponential form.

6. $y = \log_5(2x + 1)$ $\boxed{5^y = 2x+1}$

25

8. $27 = 3^x$

9. $32 = 2^{(2x + 1)}$

10 · 3 Logarithmic Functions

1. Given $y = 3^x$, the inverse of $y = 3^x$ is ___________. Since it is not convenient to solve for y by conventional methods, the logarithmic function $y = \log_a x$ is defined. In this case, $x = 3^y$ could be written _______________.

2. If the logarithmic function $y = \log_a x$ is written in the form $y = \log_e x$, this is known as the _________________________________ and is written _______________.

Rewrite the following in logarithmic form.

3. $y = 2^e$

4. $N = .035^{(x + 1)}$

5. $K = e^{1.5}$

Rewrite the following in exponential form.

6. 25 $y = \log_5 (2x + 1)$

7. $\log_x 3 = 2.6$ $\boxed{x^{2.6} = 3}$

8. $m = \ln 3$ $\boxed{e^m = 3}$

Sketch the graph of the following function.

9. $f(x) = \log_2 2x$

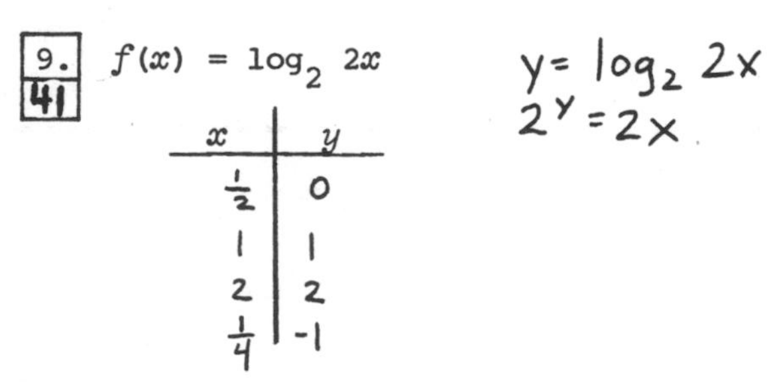

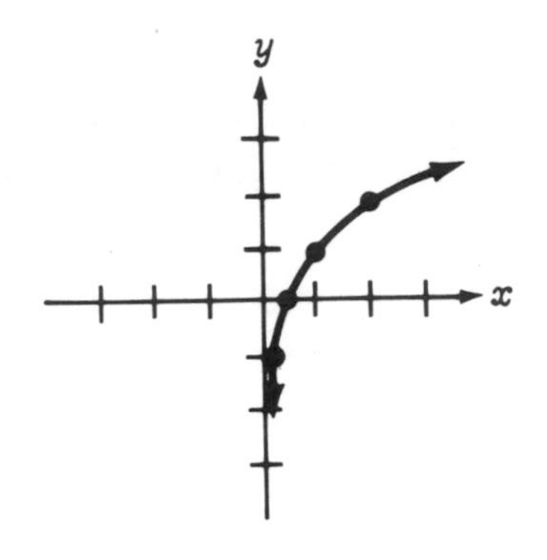

10·4 Properties of Logarithms

Write the correct choice in the blank provided.

e	1. $\log_2 (13/5)$	a.	$1/2 \log_2 x - 1/2 \log_2 y$
b	2. $\log_5 13^2$	b.	$2 \log_5 13$
i	3. $\log_3 45$	c.	$\log_8 12 - \log_8 3$
f	4. $\log_3 12$	d.	$2 \log_e 13$
c	5. $\log_8 4$	e.	$\log_2 13 - \log_2 5$
h	6. $\log_2 (x + 1)^4$	f.	$(\log_{10} 12)/(\log_{10} 3)$
d	7. $\ln 13^2$	g.	5
a	8. $\log_2 \sqrt{x/y}$	h.	$4 \log_2 (x + 1)$
g	9. $\log_3 3^5$	i.	$\log_3 9 + \log_3 5$

7. $\log_x 3 = 2.6$

8. $m = \ln 3$

Sketch the graph of the following function.

9. $f(x) = \log_2 2x$

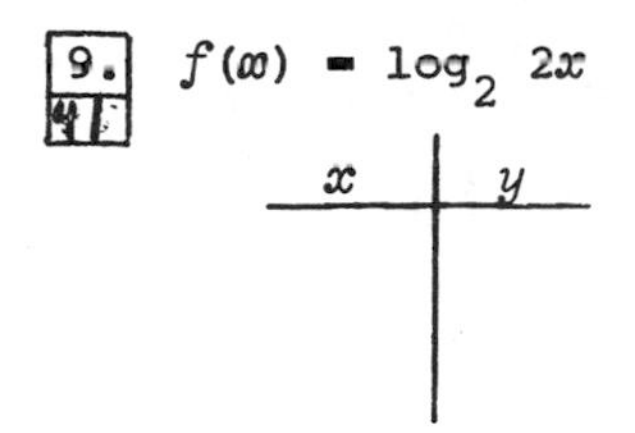

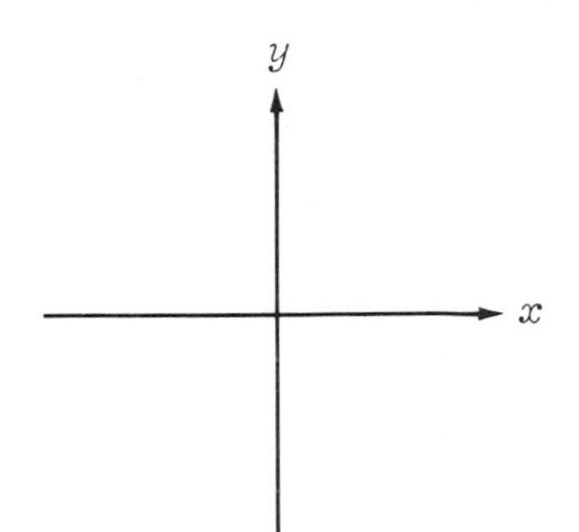

10·4 Properties of Logarithms

Write the correct choice in the blank provided.

_____	1. $\log_2 (13/5)$	a. $1/2 \log_2 x - 1/2 \log_2 y$
_____	2. $\log_5 13^2$	b. $2 \log_5 13$
_____	3. $\log_3 45$	c. $\log_8 12 - \log_8 3$
_____	4. $\log_3 12$	d. $2 \log_e 13$
_____	5. $\log_8 4$	e. $\log_2 13 - \log_2 5$
_____	6. $\log_2 (x + 1)^4$	f. $(\log_{10} 12)/(\log_{10} 3)$
_____	7. $\ln 13^2$	g. 5
_____	8. $\log_2 \sqrt{x/y}$	h. $4 \log_2 (x + 1)$
_____	9. $\log_3 3^5$	i. $\log_3 9 + \log_3 5$

Simplify each of the following.

10. $\log_b (27a^3)$

$\log_b 27 + \log_b a^3$

$\boxed{\log_b 27 + 3\log_b a}$

11. $\log_b \sqrt{3a^2b}$

$\log_b (3a^2b)^{\frac{1}{2}}$

$\frac{1}{2} \log_b 3a^2b$

$\frac{1}{2} \log_b 3 + \frac{1}{2} \log_b a^2 + \frac{1}{2} \log_b b$

$\boxed{\frac{1}{2} \log_b 3 + \log_b a + \frac{1}{2}}$

12. $\log_5 (5^6)$

$6 \log_5 5$

$6(1)$

$\boxed{6}$

13. $\log_b \sqrt[3]{(x^2/y)}$

$\log_b \left(\frac{x^2}{y}\right)^{\frac{1}{3}}$

$\frac{1}{3}(2 \log_b x - \log_b y)$

$\boxed{\frac{2}{3} \log_b x - \frac{1}{3} \log_b y}$

14. $\log_3 (1/9)$

$\log_3 1 - \log_3 9$

$\log_3 1 - \log_3 3^2$

$\boxed{\log_3 1 - 2}$

15. $\log_4 (2/3)^5$

$5 \log_4 \frac{2}{3}$

$5[\log_4 2 - \log_4 3]$

$5 \log_4 4^{\frac{1}{2}} - 5 \log_4 3$

$(5)(\frac{1}{2}) \log_4 4 - 5 \log_4 3$

$5(\frac{1}{2})(1) - 5 \log_4 3$

$\boxed{\frac{5}{2} - 5 \log_4 3}$

Simplify each of the following.

10. $\log_b (27a^3)$

11. $\log_b \sqrt{3a^2b}$

12. $\log_5 (5^6)$

13. $\log_b \sqrt[3]{(x^2/y)}$

14. $\log_3 (1/9)$

15. $\log_4 (2/3)^5$

Given $\log_b 3 = 1.4$, $\log_b 5 = 2$, $\log_b 7 = 2.4$, find the following.

16. [45] $\log_b 45 = \log_b(3^2 \cdot 5)$

$= 2\log_b 3 + \log_b 5$

$= 2(1.4) + 2$

$= \boxed{4.8}$

17. $\log_b (49/3)$

$\log_b\left(\frac{7^2}{3}\right) =$

$= 2\log_b 7 - \log_b 3$

$= 2(2.4) - 1.4$

$= \boxed{3.4}$

10.5 Logarithmic and Exponential Equations

Solve using logarithms.

1. $25 = 4^x$

$\log 25 = x \log 4$

$1.39794 = x(.602059991)$

$\boxed{x = 2.3219}$

2. [13] $200 = 125(2.45)^{2t}$

$\log 200 = \log 125 + 2t \log 2.45$

$\log 200 - \log 125 = 2t \log 2.45$

$\frac{\log 200 - \log 125}{\log 2.45} = 2t$

$.524506 = 2t$

$\boxed{.262253 = t}$

3. $5 = 20(3.15)^{(2t - 1)}$

$\log 5 = \log 20 + (2t-1)\log 3.15$

$\frac{\log 5 - \log 20}{\log 3.15} = 2t - 1$

$-1.2082 = 2t - 1$

$-.2082 = 2t$

$\boxed{-.1041 = t}$

4. $\log_x 8 = 1/2$

$x^{\frac{1}{2}} = 8$ OR $\frac{1}{2}\log x = \log 8$

$x^{\frac{1}{2}} = 64^{\frac{1}{2}}$ $\qquad \frac{1}{2}\log x = .9030$

$x = 64$ $\qquad \log x = 1.8061799$

$10^{1.8061799} = x$

$\boxed{64 = x}$

5. $\log_8(x + 1) = -2/3$

$8^{-\frac{2}{3}} = x + 1$ $\qquad (8^{\frac{1}{3}})^{-2} = x + 1$

$2^{-2} = x + 1$ $\qquad \frac{1}{4} = x + 1$

$\boxed{-\frac{3}{4} = x}$

6. $\log_5 (1/125) = x$

$5^x = \frac{1}{125}$ $\qquad 5^x = \left(\frac{1}{5}\right)^3$

$5^x = 5^{-3}$ $\qquad \boxed{x = -3}$

Given $\log_b 3 = 1.4$, $\log_b 5 = 2$, $\log_b 7 = 2.4$, find the following.

16. (45) $\log_b 45$

17. $\log_b (49/3)$

10.5 Logarithmic and Exponential Equations

Solve using logarithms.

1. $25 = 4^x$

2. (13) $200 = 125(2.45)^{2t}$

3. $5 = 20(3.15)^{(2t - 1)}$

4. $\log_x 8 = 1/2$

5. $\log_8 (x + 1) = -2/3$

6. $\log_5 (1/125) = x$

7. [27] $\log_3 x + \log_3(x + 6) = 3$

$\log_3 x(x+6) = 3$

$3^3 = x^2 + 6x$

$27 = x^2 + 6x$

$0 = x^2 + 6x - 27$

$0 = (x+9)(x-3)$

$x+9=0$ OR $x-3=0$

$x = -9$ $\boxed{x=3}$

$\log_3 -9$ is impossible

8. $\log(1 - 3x) - 2\log x = 1$

$\log \frac{1-3x}{x^2} = 1$

$10^1 = \frac{1-3x}{x^2}$

$10x^2 = 1-3x$

$10x^2 + 3x - 1 = 0$

$(5x-1)(2x+1) = 0$

$5x-1=0$ OR $2x+1=0$

$\boxed{x = \frac{1}{5}}$ $x = -\frac{1}{2}$

$2\log -\frac{1}{2}$ is impossible.

9. [35] $\log(x + 1) = \log(x + 9) + \log(x)$

$\log(x+1) = \log x(x+9)$

$x+1 = x^2 + 9x$

$0 = x^2 + 8x - 1$

$x = \frac{-8 \pm \sqrt{64+4}}{2} = \frac{-8 \pm \sqrt{68}}{2}$

$\boxed{x = .1231}$ OR $x = -8.123$

$\log -8.123$ is impossible

7. 27 $\log_3 x + \log_3(x + 6) = 3$

8. $\log(1 - 3x) - 2\log x = 1$

9. 35 $\log(x + 1) = \log(x + 9) + \log(x)$

Chapter 10 Self-Test

Graph each equation and its inverse on the same coordinate system.

1. $f(x) = 2x^2 - 32$

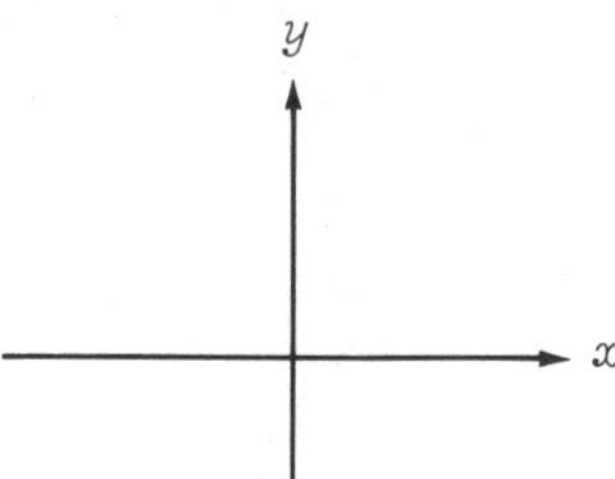

2. $f(x) = |x - 2|$

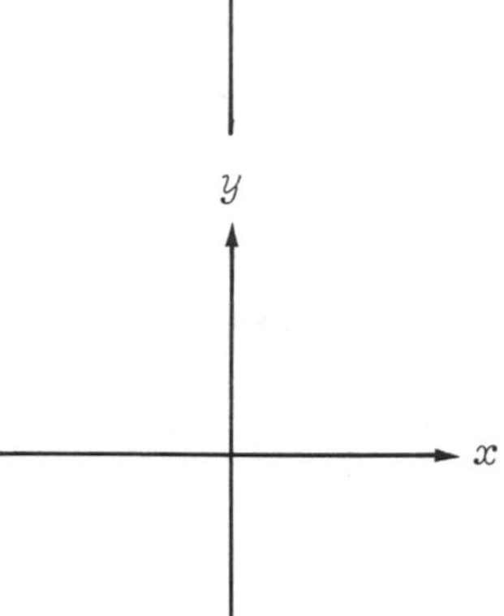

Graph each equation.

3. $f(x) = 5 + 2^x$

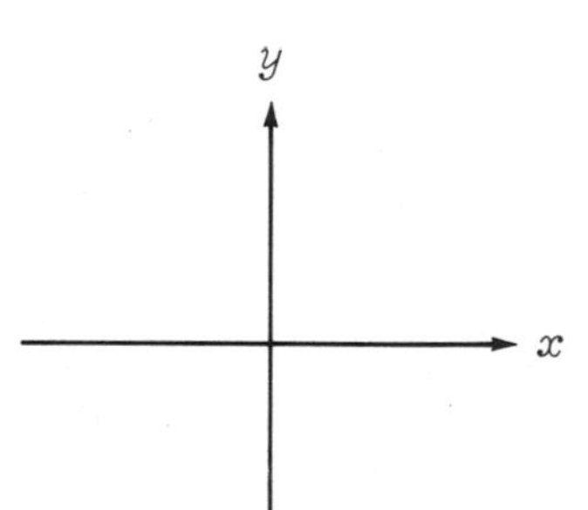

4. $f(x) = 2^{(-x)}$

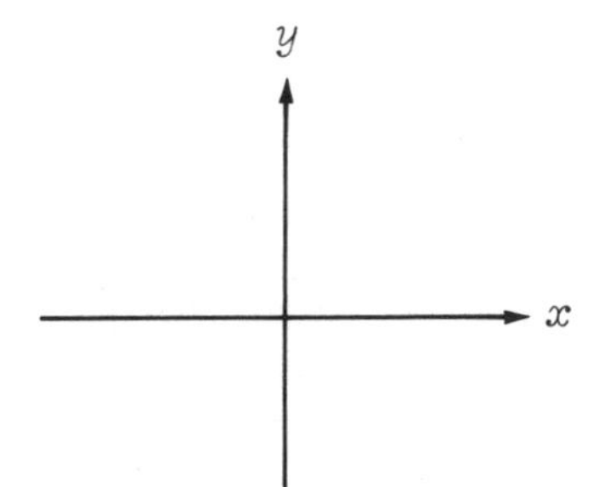

5. $f(x) = \log_3 2x$

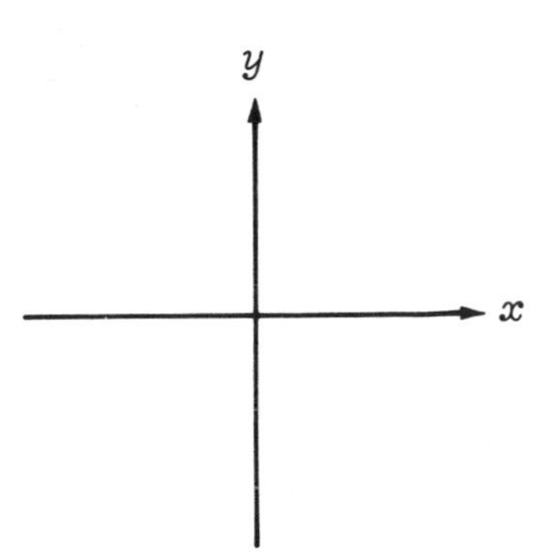

Simplify each of the following.

6. $\log_4 16^{(-2)}$

7. $\log[20(3.15)(1/2)]^{(-4)}$

Solve each equation by use of logarithms.

8. $\log_2 x + \log_2 (x - 3) = 2$

9. $8 = 15(3.12)^{(2x)}$

C H A P T E R

11 Sequences and Series

11 · 1 Sequences and Series

1. A finite sequence is a function whose domain is $\{1,2,3,4,\ldots,k\}$ for some natural number k. If the domain is $\{1,2,3,4,\ldots\}$, then the function is called an infinite sequence.

2. The elements of the range of the function are called terms. The list of terms stated in order is called a sequence.

3. The sum of consecutive terms of a sequence is called a series. If the sequence is infinite, the series is called an infinite series. The notation S_n denotes the sum of the first n terms.

4. The notation $\sum_{i=1}^{n} a_i$ is read "the summation of a_i as i assumes values from 1 to n."

5. Determine the sixth term of the sequence with the following general term. $a_n = 3^n + 5$ $\quad a_6 = 3^6 + 5 = \boxed{734}$

C H A P T E R

11 Sequences and Series

11 · 1 Sequences and Series

1. A ____________________ is a function whose domain is $\{1,2,3,4,\ldots,k\}$ for some natural number k. If the domain is $\{1,2,3,4,\ldots\}$, then the function is called an ____________________.

2. The elements of the range of the function are called __________. The list of terms stated in order is called a ____________.

3. The sum of consecutive terms of a sequence is called a __________. If the sequence is infinite, the series is called an ____________ __________. The notation S_n denotes the sum of the first ________.

4. The notation __________ is read "the summation of a_i as i assumes values from 1 to n."

5. Determine the sixth term of the sequence with the following general term. $a_n = 3^n + 5$

6. 23 Determine the ninth term of the sequence with the following general term. $a_n = (2n)/(n^3 + 1)$

$a_9 = (2.9)/(9^3+1) = \frac{18}{730} = \boxed{\frac{9}{365}}$

7. Determine the sum of the first four terms of the sequence with the following general term. $a_n = (-1)^n(3n)$

$a_1 = (-1)^1(3\cdot1) = -3$
$a_2 = (-1)^2(3\cdot2) = 6$
$a_3 = (-1)^3(3\cdot3) = -9$
$a_4 = (-1)^4(3\cdot4) = 12$
$a_1 + a_2 + a_3 + a_4 = \boxed{6}$

8. 37 Determine the sum of the first five terms of the sequence with the following general term. $a_n = n^2 - 4$

$a_1 = 1^2 - 4 = -3$
$a_2 = 2^2 - 4 = 0$
$a_3 = 3^2 - 4 = 5$
$a_4 = 4^2 - 4 = 12$
$a_5 = 5^2 - 4 = 21$
Sum $= \boxed{35} = S_5$

Write the following series in expanded form and evaluate.

9. 51 $\sum_{i=3}^{7} 2i(i + 4)$

$2\cdot3(3+4) = 42$
$2\cdot4(4+4) = 64$
$2\cdot5(5+4) = 90$
$2\cdot6(6+4) = 120$
$2\cdot7(7+4) = 154$
SUM $= \boxed{470}$

6. 23 Determine the ninth term of the sequence with the following general term. $a_n = (2n)/(n^3 + 1)$

7. Determine the sum of the first four terms of the sequence with the following general term. $a_n = (-1)^n(3n)$

8. 37 Determine the sum of the first five terms of the sequence with the following general term. $a_n = n^2 - 4$

Write the following series in expanded form and evaluate.

9. 51 $\sum_{i=3}^{7} 2i(i + 4)$

10. $\sum_{i=1}^{6} (3i - 1)/(i^2)$

$\frac{(3\cdot1-1)}{1^2} = 2$

$\frac{(3\cdot2-1)}{2^2} = \frac{5}{4}$

$\frac{(3\cdot3-1)}{3^2} = \frac{8}{9}$

$\frac{(3\cdot4-1)}{4^2} = \frac{11}{16}$

$\frac{(3\cdot5-1)}{5^2} = \frac{14}{25}$

$\frac{(3\cdot6-1)}{6^2} = \frac{17}{36}$

SUM = 5.85861

---

11 · 2 Arithmetic Sequences and Series

1. An arithmetic sequence is a sequence in which each term after the first term can be found by adding a fixed constant to the previous term. This fixed constant is called the common difference and is denoted by d.

2. To find the general term of an arithmetic sequence with first term a_1 and common difference d, the formula is $a_n = a_1 + (n-1)d$.

3. The sum of consecutive terms of an arithmetic sequence is called an arithmetic series.

4. The formula for calculating the sum of the first n terms of an arithmetic sequence is $S_n = \frac{n}{2}(a_1 + a_n)$ or $S_n = \frac{n}{2}[2a_1 + (n-1)d]$.

Determine the indicated term of the following arithmetic sequences.

5. -7, -2, 3, 8, ... Determine the 15th term, a_{15}.

$a_n = a_1 + (n-1)d$

$a_{15} = -7 + (14)(5)$

$a_{15} = 63$

6. 1/3, 5/3, 9/3, ... Determine the 21st term, a_{21}.

$a_n = a_1 + (n-1)d$

$a_{21} = \frac{1}{3} + (20)(\frac{4}{3})$

$a_{21} = \frac{81}{3} = 27$

10. $\sum_{i=1}^{6} (3i - 1)/(i^2)$

11 · 2 Arithmetic Sequences and Series

1. An ________________________ is a sequence in which each term after the first term can be found by adding a fixed ______________ to the previous term. This fixed constant is called the ___________________ ______________ and is denoted by _____.

2. To find the general term of an arithmetic sequence with first term a_1 and common difference d, the formula is ______________________.

3. The sum of consecutive terms of an arithmetic sequence is called an _______________________.

4. The formula for calculating the sum of the first n terms of an arithmetic sequence is ____________________ or ____________________.

Determine the indicated term of the following arithmetic sequences.

5. -7, -2, 3, 8, ... Determine the 15th term, a_{15}.

6. 1/3, 5/3, 9/3, ... Determine the 21st term, a_{21}.

7. $a_1 = 3$, $d = -4$ Determine a_{35}.

$a_{35} = a_1 + (n-1)d$

$a_{35} = 3 + (34)(-4)$

$a_{35} = \boxed{-133}$

8. 25 $a_7 = 22$, $a_{12} = 37$ Determine a_1.

$a_7 = a_1(n-1)d$

$a_7 = a_1 + (6)(3)$

$22 = a_1 + 18$

$a_1 = \boxed{4}$

9. 17 $-2, -5, -8, -11, \ldots$ Determine a_n.

$a_n = a_1 + (n-1)d$

$a_n = -2 + (n-1)(-3)$

$a_n = -2 - 3n + 3$

$a_n = \boxed{1-3n}$

Determine the indicated sums of the following arithmetic sequences.

10. $a_1 = -22$, $a_6 = -2$ Determine S_6.

$a_6 - a_1 = 20$

$20 \div 5 = 4 = d$

$S_n = \frac{n}{2}[2a_1 + (n-1)d]$

$S_6 = \frac{6}{2}[2(-22) + (5)(4)]$

$S_6 = 3(-44 + 20)$

$S_6 = \boxed{-72}$

-OR-

$S_n = \frac{n}{2}(a_1 + a_n)$

$S_6 = \frac{6}{2}(-22 + -2)$

$S_6 = \boxed{-72}$

11. $a_1 = 5$, $a_2 = 6.5$ Determine S_{12}.

$d = a_2 - a_1 = 1.5$

$S_n = \frac{n}{2}[2a_1 + (n-1)d]$

$S_{12} = \frac{12}{2}[2(5) + (11)(1.5)]$

$S_{12} = \boxed{159}$

7. $a_1 = 3$, $d = -4$ Determine a_{35}.

8. 25 $a_7 = 22$, $a_{12} = 37$ Determine a_1.

9. 17 $-2, -5, -8, -11, \ldots$ Determine a_n.

Determine the indicated sums of the following arithmetic sequences.

10. $a_1 = -22$, $a_6 = -2$ Determine S_6.

11. $a_1 = 5$, $a_2 = 6.5$ Determine S_{12}.

12. 43 $\sum_{i=1}^{48} (1/2\, i - 3/4)$

$a_1 = \frac{1}{2} \cdot 1 - \frac{3}{4} = -\frac{1}{4}$

$a_{48} = \frac{1}{2} \cdot 48 - \frac{3}{4} = 23\frac{1}{4}$

$S_{48} = \frac{n}{2}(a_1 + a_n)$

$= \frac{48}{2}(-\frac{1}{4} + 23\frac{1}{4})$

$S_{48} = \boxed{552}$

-OR-

$\left.\begin{matrix} a_1 = \frac{1}{2} \cdot 1 - \frac{3}{4} = -\frac{1}{4} \\ a_2 = \frac{1}{2} \cdot 2 - \frac{3}{4} = \frac{1}{4} \end{matrix}\right\} d = \frac{1}{2}$

$S_n = \frac{n}{2}[2a_1 + (n-1)d]$

$S_{48} = \frac{48}{2}[2(-\frac{1}{4}) + (47)(\frac{1}{2})]$

$S_{48} = \boxed{552}$

---

11 · 3 Geometric Sequences and Series

1. A geometric sequence is a sequence in which each term after the first term can be found by multiplying the previous term by a fixed constant. This fixed constant is called the common ratio and is denoted by r.

2. To find the general term of a geometric sequence with first term a_1 and common ratio r, the formula is $a_n = a_1 r^{n-1}$.

3. The sum of consecutive terms of a geometric sequence is called a geometric series.

4. The formula for calculating the sum of the first n terms of a geometric sequence is $S_n = \frac{a_1 - a_1 r^n}{1-r}$, $r \neq 1$.

5. The formula for calculating the sum of the terms of an infinite geometric sequence when $|r| < 1$ is $S_\infty = \frac{a_1}{1-r}$.

12. 43

$$\sum_{i=1}^{48} (1/2\, i - 3/4)$$

11 · 3 Geometric Sequences and Series

1. A _____________ _____________ is a sequence in which each term after the first term can be found by multiplying the previous term by a fixed _____________. This fixed constant is called the _____________ _____________ and is denoted by ______.

2. To find the general term of a geometric sequence with first term a_1 and common ratio r, the formula is _______________.

3. The sum of consecutive terms of a geometric sequence is called a _____________ _____________.

4. The formula for calculating the sum of the first n terms of a geometric sequence is ___________________________.

5. The formula for calculating the sum of the terms of an infinite geometric sequence when $|r| < 1$ is _______________.

Determine the indicated terms of the following geometric sequences.

6. 1/2, 1/4, 1/8, ... Determine a_{30}. $r = \frac{1}{2}$

$a_n = a_1 r^{n-1}$

$a_{30} = (\frac{1}{2})(\frac{1}{2})^{29}$

$a_{30} = \boxed{9.313 \times 10^{-10}}$

7. $a_1 = 1/3$, $r = 1/2$ Determine a_{15}.

$a_{15} = (\frac{1}{3})(\frac{1}{2})^{14}$

$a_{15} = \boxed{.000020345}$

8. $a_1 = -20$, $r = -2$ Determine a_{12}.

$a_{12} = (-20)(-2)^{11}$

$a_{12} = \boxed{40960}$

9. $a_1 = 400$, $r = 1/4$ Determine a_n.

$a_n = \boxed{(400)(\frac{1}{4})^{n-1}}$

10. (21) 16, 12, 9, 27/4, ... Determine a_n. $r = a_2 \div a_1 = \frac{3}{4}$

$a_n = \boxed{(16)(\frac{3}{4})^{n-1}}$

Determine the indicated sums of the following geometric sequences.

11. 8, 12, 18, 27, ... Determine S_{10}.

$S_n = \frac{a_1 - a_1 r^n}{1-r}$ $S_{10} = \frac{8-8(\frac{3}{2})^{10}}{1-\frac{3}{2}} = \boxed{906.64}$

12. 1/3, -1/9, 1/27, ... Determine S_∞.

$S_\infty = \frac{a_1}{1-r}$ $S_\infty = \frac{\frac{1}{3}}{1-(-\frac{1}{3})} = \frac{\frac{1}{3}}{\frac{4}{3}} = \boxed{\frac{1}{4}}$

Determine the indicated terms of the following geometric sequences.

6. 1/2, 1/4, 1/8, ... Determine a_{30}.

7. $a_1 = 1/3$, $r = 1/2$ Determine a_{15}.

8. $a_1 = -20$, $r = -2$ Determine a_{12}.

9. $a_1 = 400$, $r = 1/4$ Determine a_n.

10. 21 16, 12, 9, 27/4, ... Determine a_n.

Determine the indicated sums of the following geometric sequences.

11. 8, 12, 18, 27, ... Determine S_{10}.

12. 1/3, -1/9, 1/27, ... Determine S_∞ .

13. $\sum_{i=1}^{\infty} (2/3)^i$

$a_1 = \left(\frac{2}{3}\right)^1 = \frac{2}{3}$, $a_2 = \left(\frac{2}{3}\right)^2 = \frac{4}{9}$ $\Big\}$ $r = a_2 \div a_1 = \boxed{\frac{2}{3}}$

$S_\infty = \frac{\frac{2}{3}}{1-\frac{2}{3}} = \frac{\frac{2}{3}}{\frac{1}{3}} = \boxed{2}$

14. $\sum_{i=1}^{9} 3^i$

$a_3 = 3^3 = 27$

$a_4 = 3^4 = 81$

$\vdots$

$a_9 = 3^9 = 19683$

$n = 7$

$S_7 = \frac{27 - 27(3)^7}{1-3} = \boxed{29511}$

---

11 · 4 The Binomial Theorem and Pascal's Triangle

1. Complete the first eight rows of Pascal's Triangle.

$(a + b)^1$	1 1
$(a + b)^2$	1 2 1
$(a + b)^3$	1 3 3 1
$(a + b)^4$	1 4 6 4 1
$(a + b)^5$	1 5 10 10 5 1
$(a + b)^6$	1 6 15 20 15 6 1
$(a + b)^7$	1 7 21 35 35 21 7 1
$(a + b)^8$	1 8 28 56 70 56 28 8 1

2. Expand $(a + b)^5$ using Pascal's Triangle for the coefficients.

$\underline{1}\,a^5 + \underline{5}\,a^4b + \underline{10}\,a^3b^2 + \underline{10}\,a^2b^3 + \underline{5}\,ab^4 + \underline{1}\,b^5$

3. Expand $(x + 2)^4$ using Pascal's Triangle for the coefficients.

$\underline{1}\,x^4 + \underline{4}\,x^3 2^1 + \underline{6}\,x^2 2^2 + \underline{4x2^3} + \underline{1 \cdot 2^4}$

$\underline{x^4} + \underline{8x^3} + \underline{24x^2} + \underline{32x} + \underline{16}$

13. $\sum_{i=1}^{\infty} (2/3)^{i}$

14. $\sum_{i=1}^{9} 3^{i}$

11·4 The Binomial Theorem and Pascal's Triangle

1. Complete the first eight rows of Pascal's Triangle.

$(a + b)^1$	1 1
$(a + b)^2$	1 2 1
$(a + b)^3$	1 3 3 1
$(a + b)^4$	
$(a + b)^5$	
$(a + b)^6$	
$(a + b)^7$	
$(a + b)^8$	

2. Expand $(a + b)^5$ using Pascal's Triangle for the coefficients.

$\_\_\_a^5 + \_\_\_a^4b + \_\_\_a^3b^2 + \_\_\_a^2b^3 + \_\_\_ab^4 + \_\_\_b^5$

3. Expand $(x + 2)^4$ using Pascal's Triangle for the coefficients.

$\_\_\_x^4 + \_\_\_x^3 2^1 + \_\_\_x^2 2^2 + \_\_\_\_\_\_\_\_ + \_\_\_\_\_\_\_\_$

$\_\_\_\_\_\_\_ + \_\_\_\_\_\_\_ + \_\_\_\_\_\_\_\_ + \_\_\_\_\_\_\_ + \_\_\_\_\_\_\_\_$

4. Expand $(x + 2)^4$ using the Binomial Theorem.

$$x^4 + \frac{4}{1!} x^3 2^1 + \frac{4 \cdot 3}{2!} x^2 2^2 + \frac{4 \cdot 3 \cdot 2}{3!} x 2^3 + \frac{4 \cdot 3 \cdot 2 \cdot 1}{4!} \cdot 2^4$$

$$x^4 + 8x^3 + 24x^2 + 32x + 16$$

4. Expand $(x + 2)^4$ using the Binomial Theorem.

$$x^4 + \frac{4}{1!} x^3 2^1 + \frac{4 \cdot 3}{2!} x^2 2^2 + \qquad\qquad +$$

Chapter 11 Self-Test

Given the following information, determine the nth term of the sequence.

1. $a_n = 2^n - 1$; find a_6

2. $a_n = (1/3)^n + 2$; find a_4

3. $a_1 = 1/4$, $r = 2$; find a_{12}

4. $a_1 = -6$, $r = 1/2$; find a_6

5. $-8, 12, -18, 27, \ldots$; find a_9

6. $-7, -11/2, -4, \ldots$; find a_{20}

Determine the sum of the first n terms of each series.

7. Arithmetic: $a_1 = -15$, $a_4 = -6$; find S_{10}

8. Arithmetic: $\sum_{i=1}^{48} (2i - 3)$

9. Geometric: 3, 1, 1/3, 1/9, ...; find S_{12}

10. Geometric: $\sum_{i=1}^{\infty} \left(\frac{1}{4}\right)^{i}$

11. Expand $(x + 1)^5$ using the Binomial Theorem.

SELF-TEST SOLUTIONS

Chapter 1

1. True
2. False
3. True
4. False
5. True
6. False
7. False
8. False
9. 9
10. 2
11. undefined
12. -9
13. -5
14. $\frac{-5}{4}$
15. 2.925
16. 1.1628
17. 6
18. 1/9
19. 3/35
20. 247/360
21. -52
22. 36
23. -3, $\sqrt{9}$, $0/5$
24. π , $-\sqrt{11}$
25. Identity for addition
26. Transitive property of equality

Chapter 2

1. $x = 8/7$

2. $x = 23/3$

3. $x \geq 2.048$

2.048

4. $-5 \leq x < 8$

-5 8

5. x = any real number

6. $\emptyset$

7. $x = -11$ or $x = 13$

8. $x = 0$ or $x = -24/5$

9. $x = 8/3$ or $x = -2$

10. $0 \leq x \leq 4$

11. $\dfrac{2A - b_1 h}{h} = b_2$

12. $\dfrac{3V}{r^2 h} = \pi$

13. 37 nickels and 25 dimes

14. 12 hours

Chapter 3

1. $-9x + 21$

2. $2x^2 + 9x - 5$

3. $-.09x^2 - .06x - .15$

4. $a^6 - 8a^3 + 16$

5. $(2/9)x^2 + (1/4)xy - (1/8)y^2$

6. $p^5 + 3p^4 - 6p^3 - p^2 + 5p - 2$

7. $4y^2(x^3 - 6x^5y^2 + 3)$

8. $(x + 4)(x - 20)$

9. prime

10. $(3x + 4y)(2x - 3y)$

11. $(3x - 1)(9x^2 + 3x + 1)$

12. $(c - d)(2a + 3b)$

13. $x = 6$ or $x = -10$

14. $x = -5$

15. 5" × 8"

16. 10" × 24"

Chapter 4

1. $(2y^5)/3$
2. $(-2x - 5)/(x + 2)$
3. $(3y)/(4x^8)$
4. $(4x^2 - 2x + 1)/(2x - 1)$
5. $(2a + 1)/(2a - 3)$
6. $(8y - 12x^2 + xy^2)/(6x^2y^2)$
7. $\dfrac{(-x + 12)}{x(x - 3)(x + 3)}$
8. $\dfrac{-4x}{x + 3}$
9. $\dfrac{3}{x}$
10. $x - 3 + \dfrac{5}{x - 5}$
11. $x = 11/30$
12. $x = -4$ or $x = 1$
13. $x = -1/2$
14. $x = -4/7$ or $x = 3$
15. $x = 1.5$ hours

Chapter 5

1. $\dfrac{x^9y^{1/2}}{343}$
2. $x^{4/3} + x^{2/3} - 12$
3. $-1/3$
4. $9x^4y^2$
5. x^3
6. $3xz^2 \sqrt[3]{2x^2y^2}$
7. no solution
8. $\sqrt[3]{\dfrac{5x}{6}}$
9. $8x\sqrt{3}$
10. $(-3\sqrt{5})/5$
11. $107 - 42\sqrt{2}$
12. $\dfrac{16 - 5\sqrt{7}}{9}$
13. $x = 19/2$
14. $x = 1$
15. $x = -1/4$
16. $8 - 2i\sqrt{5}$
17. $6/25 - (8/25)i$

Chapter 6

1.

x	y
0	6
2	0
4	2
5/2	-1/4

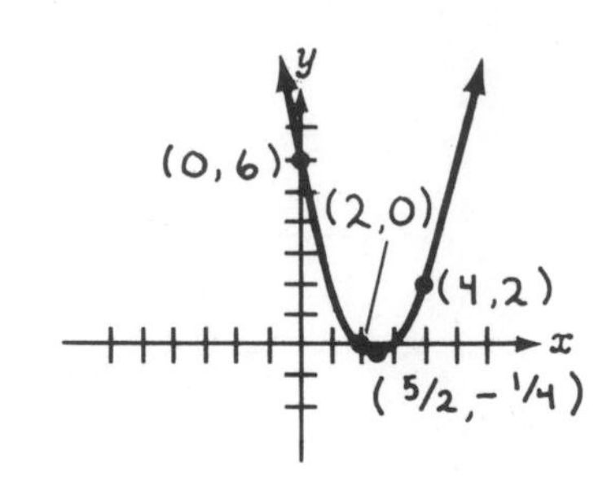

2.

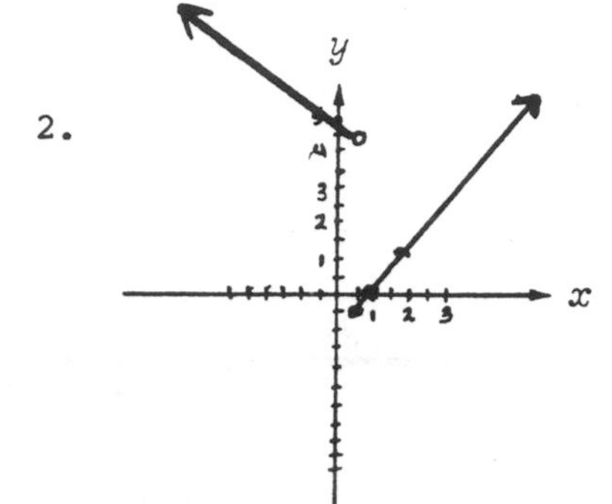

3.

x	y
2	0
0	3

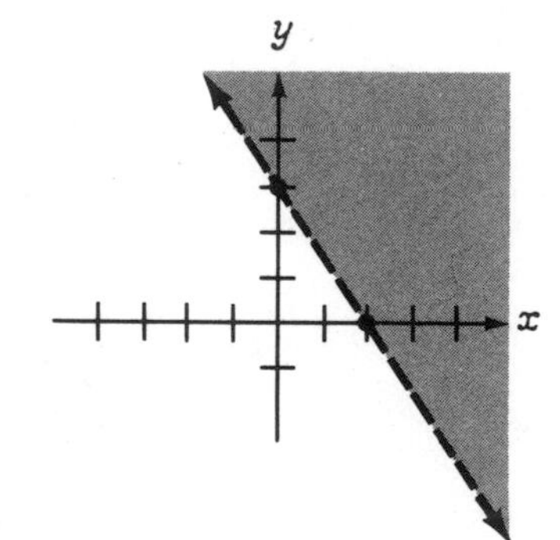

4.

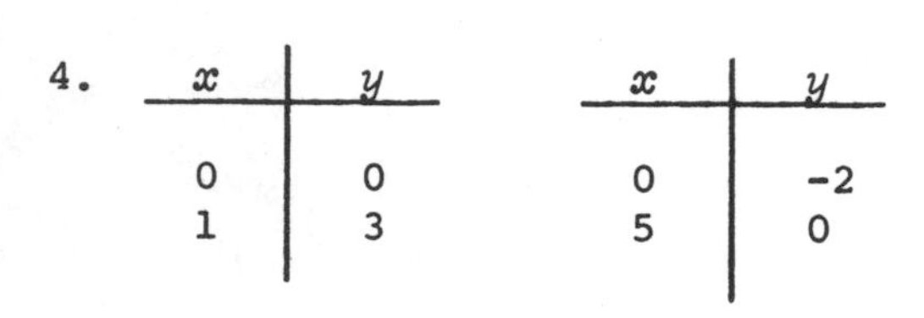

x	y
0	0
1	3

x	y
0	-2
5	0

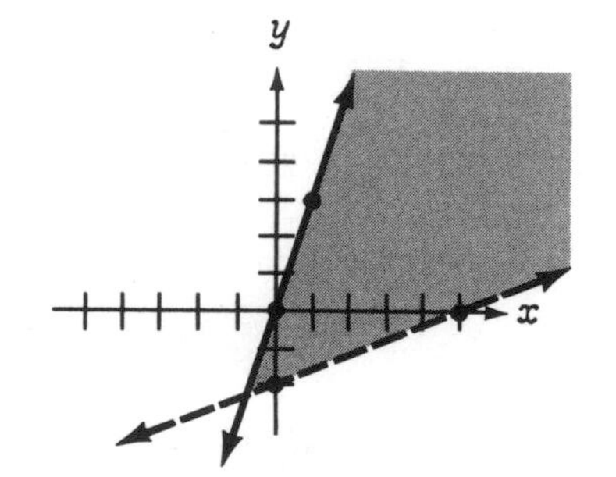

5. $3x^2 + 4x - 8$

6. -1

7. $3x - 4y = -17$

8. $3x + 16y = -7$

9. $x + 3y = -16$

10. $\sqrt{34}$

11. 266 2/3 feet

Chapter 7

1. Vertex at (0,4)

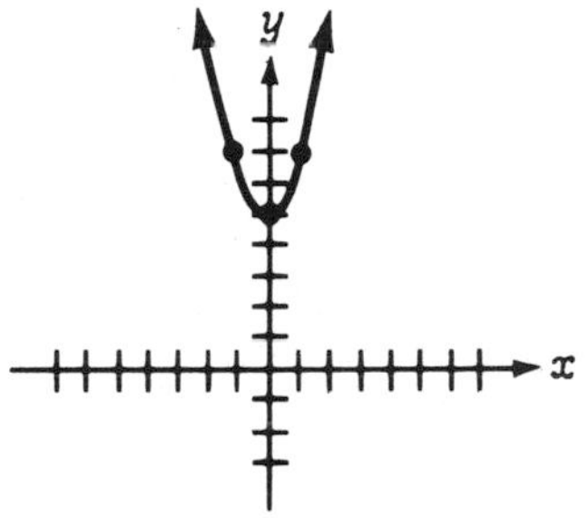

2. Vertex at (-3, -32)

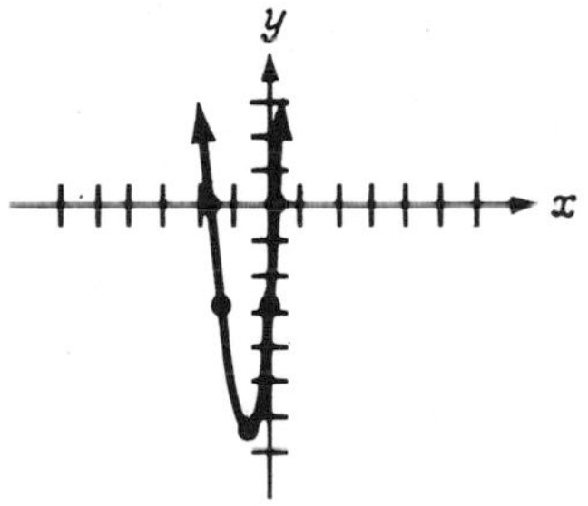

3. $x = \pm \dfrac{\sqrt{30}}{2}$

4. $x = 1 \pm 6i\sqrt{2}$

5. $x = 17$ or $x = -3$

6. $x = \dfrac{-17 \pm \sqrt{349}}{6}$

7. $x = \dfrac{-3 \pm \sqrt{185}}{8}$

8. $x = 3 \pm 2\sqrt{3}$

9. $x = 3 \pm 2\sqrt{2}$

10. $x = \pm 1\ ,\ \pm \dfrac{i\sqrt{6}}{2}$

11.

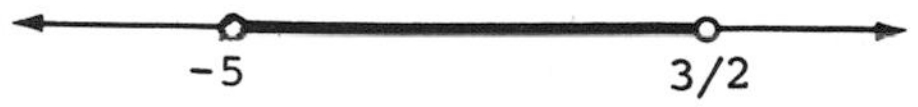

Chapter 8

1. Parabola

   Vertex at $(-13\frac{1}{2}, 3/2)$

   passing through

   (0,6) and (0,-3)

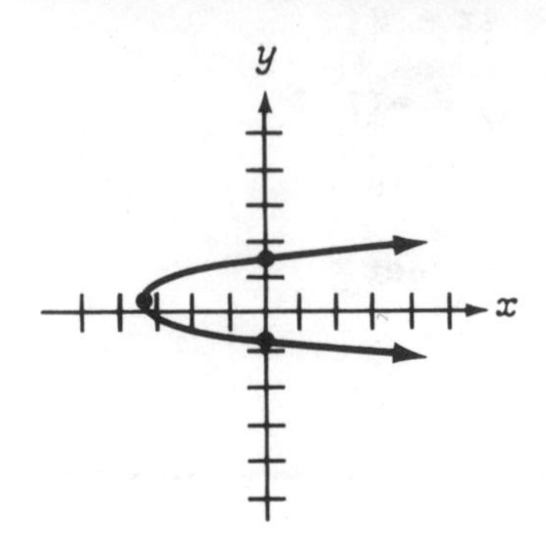

2. Circle

   Center: $(2, -\frac{1}{2})$

   Radius: 3

   $(x - 2)^2 + (y + \frac{1}{2})^2 = 9$

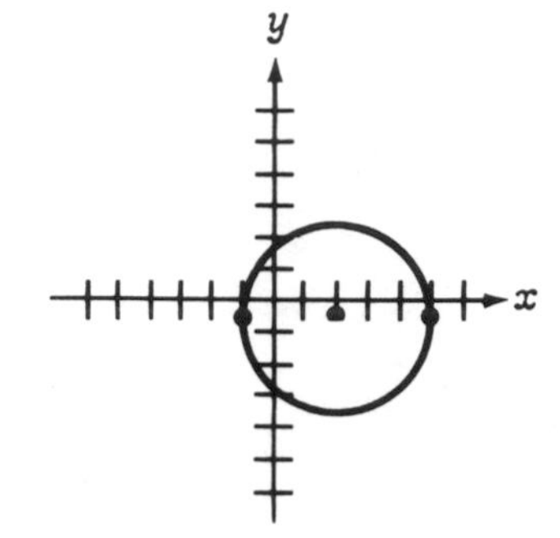

3. Ellipse

x	y
0	±2
±3	0

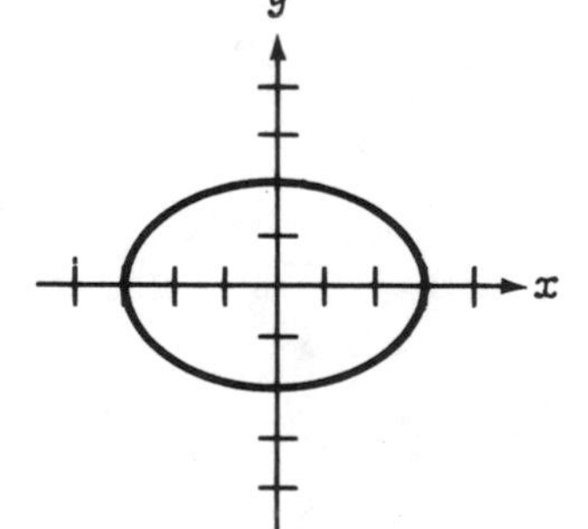

4. Hyperbola

   $$\frac{x^2}{4} - \frac{y^2}{16} = 1$$

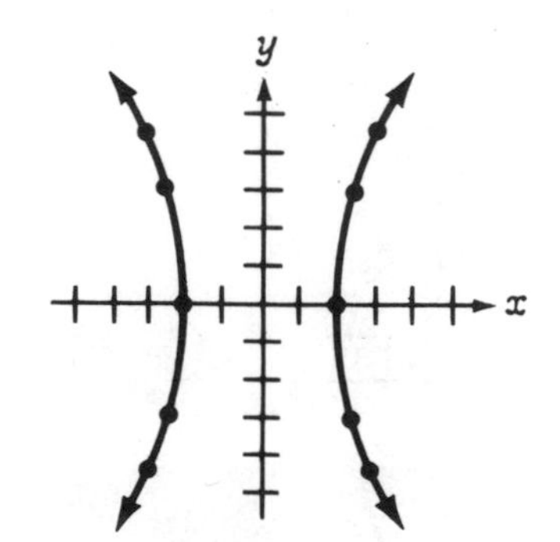

5. Parabola

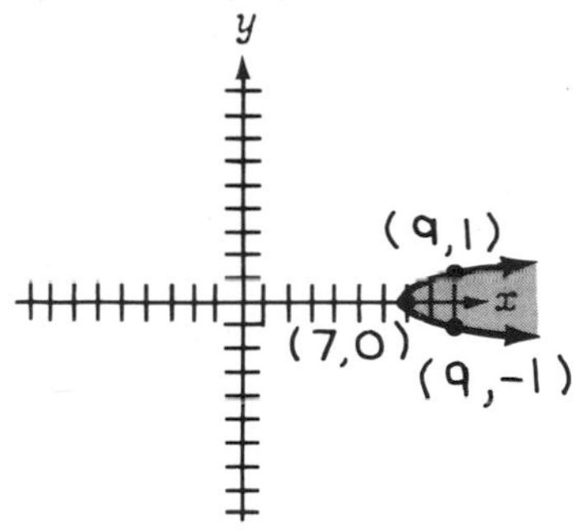

6. $(x - 2)^2 + (y - 3)^2 = 4$

Center: (2,3)

Radius: 2

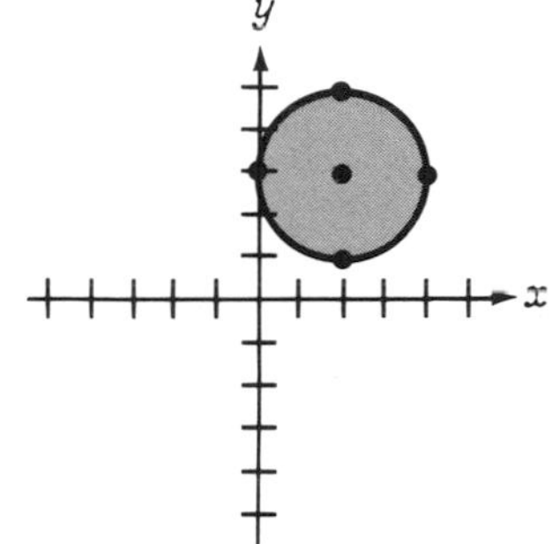

Chapter 9

1. infinite number of solutions
2. (14/31, -13/31)
3. no solution
4. (3, 2/3, -2)
5. (18/5, -1/5) or (-2, -3)
6. $(-\sqrt{5}, -4)$ or $(\sqrt{5}, -4)$
7. (5, 2)
8. (1, -4, 3)
9. \$4000 @ 7%, \$6000 @ 4%

Chapter 10

1.

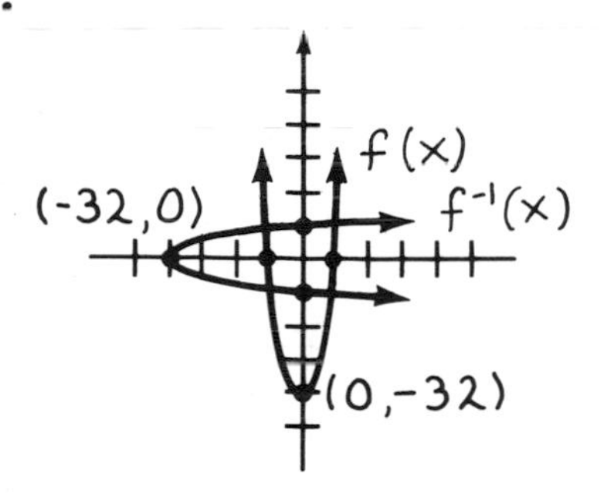

2.

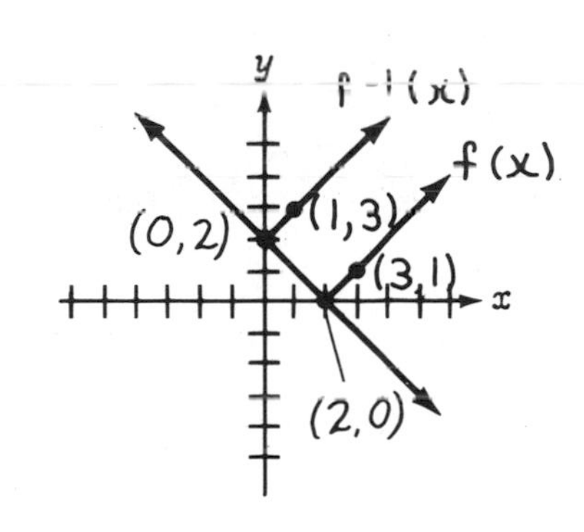

3.

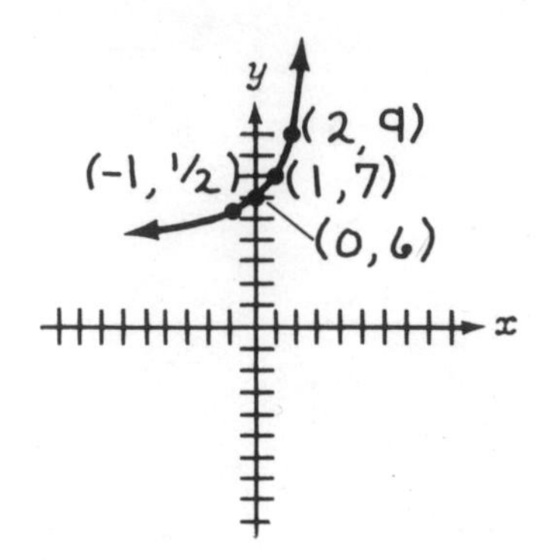

4.

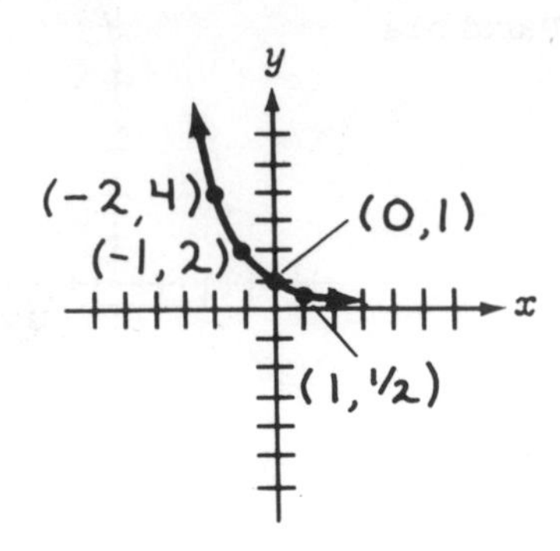

5.

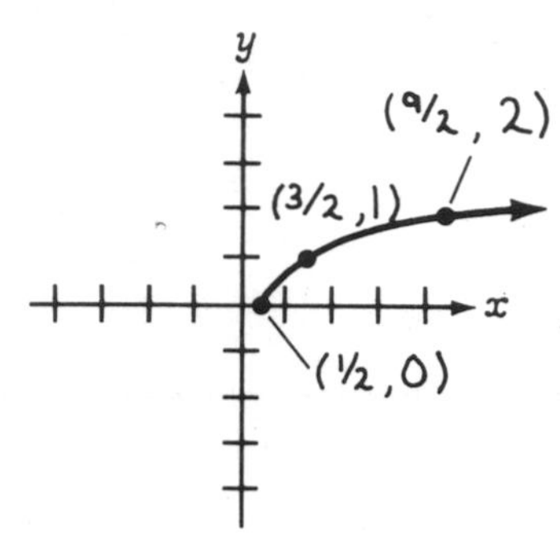

6. $x = -4$

7. -5.9932

8. $x = 4$

9. $x \doteq -.27$

Chapter 11

1. $a_6 = 63$

2. $a_4 = 2\ 1/81$

3. $a_{12} = 512$

4. $a_6 = -.1875$

5. $r = -3/2$, $a_9 = -205.03125$

6. $d = 3/2$, $a_{20} = 21.5$

7. $d = 3$, $S_{10} = -15$

8. $S_{48} = 2208$

9. $r = 1/3$, $S_{12} = 4.49999$

10. $r = 1/4$, $S_\infty = 1/3$

11. $x^5 + 5x^4 + 10x^3 + 10x^2 + 5x + 1$